W0018839

An Introduction to Traffic Flow Theory

An Introduction to Traffic Flow Theory

Editor

Shaithis Orlov

An Introduction to Traffic Flow Theory

Edited by **Shaithis Orlov**

Printed in 2017

ISBN: 978-1-68117-108-1
Library of Congress Control Number: 2015951139

© 2016 by
SCITUS Academics LLC,
616, Corporate Way, Suite 2, 4766,
Valley Cottage, NY 10989

www.scitusacademics.com

Notice

Preface

Transportation is generally concerned with the efficient, safe, and sustainable movement of people and goods. Transportation engineers work on various aspects of the five stages essential in the life cycle of a transportation facility: planning, designing, building, operating, and maintaining. In the planning stage, we typically forecast traffic demands for a future year/analysis period, perform a preliminary evaluation of alternative solutions, or identify priorities for system improvements.

This text provides a comprehensive and concise treatment of the topic of traffic flow theory and includes several topics relevant to today's highway transportation system. It provides the fundamental principles of traffic flow theory as well as applications of those principles for evaluating specific types of facilities (freeways, intersections, etc.). Newer concepts of Intelligent transportation systems (ITS) and their potential impact on traffic flow are discussed. State-of-the-art in traffic flow research and microscopic traffic analysis and traffic simulation have significantly advanced and are also discussed in this text.

This textbook is meant for use in advanced undergraduate/graduate level courses in traffic flow theory with prerequisites including two semesters of calculus, statistics, and an introductory course in transportation. The text would also be of interest to transportation professionals as a refresher in traffic flow theory, or as a reference.

Table of Contents

CHAPTER 1   Operational Characteristics of Mixed Traffic Flow Under
             Bi-Directional Environment Using Cellular Automaton
             1

ABSTRACT..1
INTRODUCTION..2
   Background..2
   Research Objectives 4
MODEL DEVELOPMENT 4
   Basic CA Model...4
   Cell Size Specification...5
   Evolvement Rules 6
MODEL CALIBRATION AND VALIDATION 10
NUMERICAL STUDY 11
   Model Installment...11
   Numerical Analysis 14
   Velocity-density Analysis 16
CONCLUSIONS..18
REFERENCES 19
CITATION...21

CHAPTER 2   Traffic Flow Consideration in Design of Freight
             Distribution System 23

ABSTRACT...23
INTRODUCTION...24
MODEL FORMULATION 26
MODEL SOLUTION 32
ILLUSTRATIVE EXAMPLE..37
CONCLUSION 41

REFERENCES 41

CITATION..42

CHAPTER 3   Real-Time Road Traffic Anomaly Detection............ 43

ABSTRACT..43

INTRODUCTION...44

METHODOLOGY 45

    Section Mutual Influence 45

    Accident Detection Strategy 46

    Incident-Influence Traffic Data 48

    Real-Time Accident Detection...49

    Accident Probability 50

    Smoothed Parameter Optimization...51

PERFORMANCE ANALYSIS...52

SIMULATION RESULTS 54

CONCLUSION 57

REFERENCES 58

CITATION..59

CHAPTER 4   Motion Planning of Autonomous Vehicles on a Dual
                  Carriageway without Speed Lanes......................... 61

ABSTRACT..61

INTRODUCTION...61

RELATED WORKS 65

MOTION PLANNING WITHOUT SPEED LANES 68

SINGLE-LANE OVERTAKING 71

    General Travel 74

    Cancelling Single-Lane Overtaking...77

    Completing Single-Lane Overtaking..79

    Algorithm 1:..79

RESULTS 85

CONCLUSIONS 90

ACKNOWLEDGMENTS 91

AUTHOR CONTRIBUTIONS 91

REFERENCES 91

CITATION..94

CHAPTER 5   Traffic Measurement on Multiple Drive Lanes with
              Wireless Ultrasonic Sensors................................ 95

ABSTRACT..95
INTRODUCTION..95
OVERVIEW OF ULTRASONIC DETECTION SYSTEMS 97
DEVELOPMENT OF A LATERAL SCANNING-BASED ULTRASONIC
SENSING SYSTEM..99
     Architecture 99
     Vehicle Detection Algorithm..101
     Noise Filtering 103
     Wireless Ultrasonic Sensor Mote (WUSM)....................107
EXPERIMENTS AND RESULTS 108
     Experimental Environment 108
     Vehicle Detection Results 109
     Noise Filtering Results According to Window Mask Size 112
CONCLUSIONS 113
AUTHOR CONTRIBUTIONS 114
REFERENCES 114
CITATION...115

CHAPTER 6   An Analysis on Efficiency and Equity of Fixed-Time
              Ramp Metering.. 117

ABSTRACT..117
INTRODUCTION..118
EFFICIENCY AND EQUITY CONCEPTS 119
     Approaches to Equity in Transportation 120
     Equity Measures and Indicators 120
TRAFFIC CONTROL STRATEGIES 122
     Ramp Metering..123
     Fixed-Time Ramp Metering 124
TRAFFIC MICRO-SIMULATION MODELING 126
     Study Site and Data 126
     Calibration and Simulation 127
EFFICIENCY AND EQUITY PERFORMANCE OF TRAFFIC CONTROL
STRATEGIES...129
     Efficiency Performance 129
     Equity Performance 130
CONCLUSIONS 131

ACKNOWLEDGEMENTS 132
REFERENCES 133
CITATION...136

CHAPTER 7   An Approach to an Intersection Traffic Delay Study
                 Based on Shift-Share Analysis............................ 137

ABSTRACT...137
INTRODUCTION..138
LITERATURE REVIEW 139
    Intersection Delay Models 140
    Right-Turn Delay Models 142
STRUCTURED DELAY ANALYSIS BASED ON SHIFT-SHARE
ANALYSIS 148
    The Premise 148
CASE STUDY..150
CONCLUSIONS..152
ACKNOWLEDGMENTS 152
AUTHOR CONTRIBUTIONS 153
REFERENCES 153
CITATION...154

CHAPTER 8   Using Multi-Attribute Decision Factors for a Modified
                 All-or-Nothing Traffic Assignment........................ 156

INTRODUCTION..156
LITERATURE REVIEW 158
MODEL DEVELOPMENT 160
    Approach 160
    Components..161
CASE STUDY..164
    System Optimum 164
    Key Attributes 165
    Scenario Analyses 167
RESULTS AND IMPLICATIONS 168
    Effects of Capacity 168
    Effects of Congestion 172
    Effects of Packet Size 172
    Discussion 172
CONCLUSIONS..174

Contents

ACKNOWLEDGMENTS 175
AUTHOR CONTRIBUTIONS 175
REFERENCES 175
CITAION 178

CHAPTER 9   Characterization of Black Spot Zones For Vulnerable Road Users In São Paulo (Brazil) And Rome (Italy). 179

ABSTRACT...179
   Background 182
   Traffic Accident Research 182
   VRU Traffic Accidents..185
   Spatial Analysis of VRU Accidents 186
STUDY AREAS..188
DESCRIPTION OF DATA AND METHODS...189
   São Paulo..189
   Rome 189
   Methodology 190
RESULTS 192
   Kernel Density Estimator..192
   Statistical Analysis 195
CONCLUSIONS 201
FINAL CONSIDERATIONS 202
ACKNOWLEDGMENTS 204
AUTHOR CONTRIBUTIONS 204
CONFLICTS OF INTEREST..204
REFERENCES 204
CITATION..211

CHAPTER 10   Comprehensive Assessment on Sustainable Development of Highway Transportation Capacity Based on Entropy Weight and TOPSIS................... 213

ABSTRACT...213
INTRODUCTION..214
ENTROPY WEIGHT MODEL 215
   Data Standardization...216
   Determination of Entropy Weight...217
TOPSIS MODEL 218

APPLICATION IN HIGHWAY TRANSPORTATION DEVELOPMENT
ASSESSMENT...219
CONCLUSIONS..222
ACKNOWLEDGMENTS 223
AUTHOR CONTRIBUTIONS 223
CONFLICTS OF INTEREST...223
REFERENCES 224
CITATION...225

CHAPTER 11   Reliable Freestanding Position-Based Routing In
               Highway Scenarios.................................... 227

ABSTRACT..227
INTRODUCTION...228
RELATED WORK 231
FREESTANDING POSITION-BASED ROUTING PROTOCOL.............234
    Radio Channel 235
    Beacon Rate..237
    Next-Hop Selection Algorithm 237
    Location Service 240
    Data Dissemination 244
    FPBR State Machine and Flow Chart 245
PERFORMANCE EVALUATION..246
    Benchmark Routing Protocol 247
    Evaluation Setup 248
    Metrics 250
    Results and Analysis 251
CONCLUSIONS AND FUTURE WORK 261
REFERENCES 262
CITATION...267

Index 269

CHAPTER 1

Operational Characteristics of Mixed Traffic Flow Under Bi-Directional Environment Using Cellular Automaton

Zhenke Luo[1], Yue Liu[1], Chen Guo[2]

[1]Department of Civil and Environmental Engineering, University of Wisconsin-Milwaukee, Milwaukee, Wisconsin, USA
[2]School of Control Science and Engineering, Shandong University, Jinan, Shandong, China

ABSTRACT

Mixed traffic flow composed of autos and non-autos widely exists in developing countries and areas. To investigate the operational characteristics of the mixed traffic flow consisting of vehicles in different types (large vehicles, cas, and bicycles), we develop a cellular automaton model to replicate the travel behaviors on a bi-directional road segment with respect to the physical and mechanic ftures of différent vehicle types. By implementing the eesential parameters calibrated through the field data collection, a numerical study is carried out considering the variation in volume, density, and velocity with different compositions of mixed traffic flows. The primary findings include: the average velocity of traffic flow and total volume decrease 60% and 30% after incorporating 10% bicycles, respectively; the phenomenon of double-summit in terms of the total volume appears when the proportion of bicycle is beyond 60%; the maximal total volume starts to recover when the proportion of bicycle is higher than 10 %.

INTRODUCTION

Background

Mixed traffic flow consisting of autos and non-autos widely exists in many developing countries and areas, such as China, India and Indonesia (Khan and Maini 1999), especially at road segments without median separation. The discrepancies in the operational characteristics of different vehicles and their interactions and interferences play an important role in affecting the traffic operational efficiencies (e. g. throughput, speed, volume, etc.). To capture and replicate such behaviors, many researchers have made aftempts on developing various types of methods, tools, and models to better understand such operational characteristic.

In review of the literature, early efforts tackling with traffic flow modeling primarily apply statistical methods to explain the fundamental relations between flow, density and speed. One pioneering work illustrated the potential of applying Poisson distribution in explaining traffic operational characteristics (Kinzer 1993). Adams published the statistical result considering the input of traffic flow as random series. Greenshields et al. (1947) used Poisson distribution in investigating the traffic flow across intersections.

In the 1950s-1970s, more researchers have developed advanced models gaining a better description of traffic flow instead of performing statistical analysis in early research. For example, the car following theory (Chander et al., 1958 and Herman et al., 1959) focused on interactions between the leading vehicles and following vehicles. Lighthill and Whitham (1955) demonstrated that the moving of traffic flow is similar to other phenomena in the natural world such as ocean waves, avalanches, and debris flows. By incorporating the fluid mechanic theory, FREFLO (Payne, 1971 and Payne, 1979) was developed and widely used in real world practices.

Recently, the advance of computational technology has facilitated the exploration of more sophisticated microscopic traffic flow models that are able to successfully capture the behaviors of individual vehicles and pedestrians with respect to various influential factors, such as types of vehicles, weathers, facility types, and control methods. Cellular automaton (CA) is one of the most prevailing and successful microscopic models. It was first applied in the transportation field to simulate car movements including lane changing, turning, queuing, acceleration, and deceleration in the road network, and the results demonstrated its ability to capture the phenomena of macroscopic models while in the meantime reproduce the mechanics of microscopic models (Cremer and Ludwig 1986). Nagel and Schreckenberg (1992) extended the cellular automaton model by setting more traffic evolvement rules to capture more realities. Fukui and Ishibashi (1996) considered that the operational speed of an individual vehicle is not only influenced by its leading vehicle, but also by the density of the neighboring environment. Foulaadvand and Belbasi (2007) studied vehicular traffic flow at an un-signalized intersection by a cellular automaton model and validated the model characteristics by a mean-field approach and extensive simulations. Ruskin and Wang (2007) studied un-signalized intersections by introducing the concept of acceptable headway.

In modeling mixed traffic flow using cellular automaton, Gundaliya et al. (2008) developed the models with multiple cell occupancy, reflecting sizes and shapes of different vehicles to reproduce the macroscopic properties of heterogeneous traffic typical of Indian cities. Jiang et al. (2004) modelled the "in bulk" movement of traffic flow consisting of bicycles. Whereas, Vasic and Ruskin (2012) modeled the mixed traffic flow in which bicycles sparsely spread in a one dimensional single lane environment. Meng et al. (2007) incorporated motorcycles into the traffic to investigate the interrelation between different vehicle types and the impact on traffic operations. Xie et al. (2009) used CA model to investigate mixed traffic flows at un-signalized intersections and suggested that the velocity difference between different types of vehicles is an important factor reflecting travel behaviors. Zhao and Gao (2005) described mixed traffic flow by combining the NaSch model (Nagel and Schreckenberg 1992) and the Burger cellular automata (BCA) model (Nishinari and Takahashi 1998), and investigated the mixed traffic system near a bus stop.

The problem to be addressed in this paper is the lack of a comprehensive model describing the traffic condition associated with mixed traffic flow at bi-direction road segments. Under a bi-directional environment without median separation, the under estimation of the impacts caused by the opposing flow at the other lane and the differences underlying lane-changing behavior will limit the model's applicability and result in simulation results far away from the reality.

Research Objectives

To contend with the above problems, the objectives of this study are to investigate the operational characteristics of mixed traffic flow consisting of autos and bicycles at bi-directional road segments without the setting of exclusive lanes and road medians. More specifically, we will determine the representation of mixed traffic flow in a two-lane bi-directional environment; develop the evolvement rules for the CA model considering the physical and mechanic characteristics of different types of vehicles under a bi-directional environment; investigate the operational characteristics of mixed traffic flow by using the proposed model.

MODEL DEVELOPMENT

Basic CA Model

Our model is based on the NaSch model which is defined on a one-dimensional array of L "cells" under open or periodic boundary conditions and each "cells" may either be occupied by at most one vehicle or be empty as shown in Fig. 1. Suppose x_n and v_n denote the location and the speed of vehicle n, respectively, and v_{max} represents the maximum velocity of a vehicle. $d_n = x_{n+1} - x_n - 1$ is the distance from the vehicle n to its front vehicle. Then each vehicle can move with an integer velocity. An update of vehicle state in the CA model involves with the following four consecutive steps.

Step 1:
Acceleration, if $v_n < v_{max}$, then increase the velocity of vehicle n by one unit, $v_n = \min\{v_n + 1, v_{max}\}$.

Step 2:
Deceleration, if $d_n < v_n$, then the speed of vehicle n is decreased to d_n, i. e. $v_n = \min\{v_n, d_n\}$.

Step 3:
Randomization, if $v_n > 0$, then the velocity decreases by one unit with a probability p which is accounting for conditions that the velocity decreases

due to the influence of other uncertain factors, such as pedestrians, obstructions, and distractions, i. e.$v_n = \max \{v_n - 1, 0\}$.

Step 4:
Movement, the vehicle updates its location with the velocity determined by Steps 1-3, i. e. $x_n = x_n + v_n$.

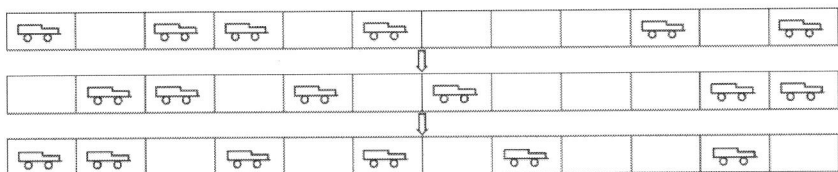

Figure 1. Illustration of NaSch model.

Cell Size Specification

To establish our model, we need to make some additional assumptions to determine the size of the "cell" aiming at achieving a convenient representation of the operations of different vehicle types in the mixed traffic flow.

Firstly, given a typical two-lane urban road with a 3-meter lane width, we need to know how many bicycles can be accommodated. According to Ren et al. (2003), we assume that one single lane can hold 3 bicycles in a row and a bicycle may share a cell with a car. A large vehicle (e. g. bus, truck) is assumed to occupy 3 m according to our observation. For the longitude space required for different types of vehicles, a bicycle is assumed to occupy a 2.75 m longitude distance in a complete traffic jam, which is the half of the distance that a car needs. While the length for a large vehicle considering the safety distance is set to be 11 m.

With the information regarding the length and width for each type of vehicle, we can determine the cell size to be 2.75 m by 3.00 m, which is able to accommodate 3 bicycles. A car takes 2 cells in length and 2/3 cell in width. While a large vehicle takes 4 cells in length, 1 cell in width. The similar concept has already been employed in the case of mixed car/truck traffic in which cars are "shorter" vehicles and trucks are "longer" vehicles (Nagel and Schreckenberg 1992). An example of mixed traffic flow representation environment at a two-lane bi-directional road segment is illustrated in Fig. 2.

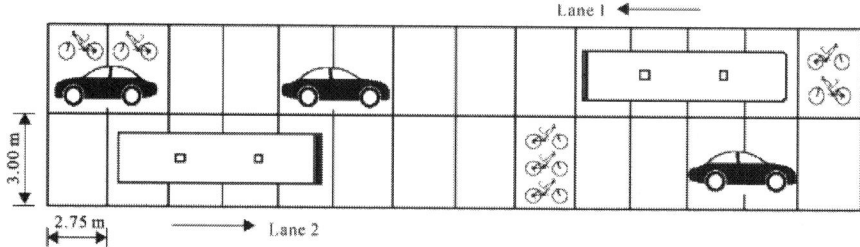

Figure 2. Representation of mixed traffic flow at a two-lane bi-directional road segment.

Evolvement Rules

The evolvement rules set for the mixed traffic flow should take into account the mechanical characteristics of different types of vehicles. For example, bicycles are smaller in size and easier to operate, therefore they have more flexibilité in acceleration, deceleration, and lane changing. However, they are associated with a relative low maximal velocity compared with other vehicles. Cars have a higher maximal velocity but a lower probability for conducting lane-changing behaviors due to the limitation in their sizes and safety considerations. In our model, car following, lane changing, and parallel operating are explicitly modeled and illustrated in Fig. 3.

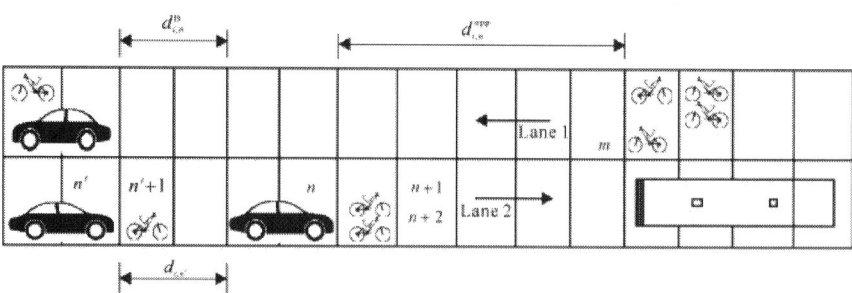

Figure 3. Illustration of driving behaviors.

For the lane changing behavior, there should be sufficient gap in the target lane with the consideration of safety. Technically, there must be sufficient intervals to meet the requirements that

$$d_{i,n}^{B} \geqslant gap_i^{B}$$
$$d_{i,n}^{opp} \geqslant gap_i^{opp}$$

where $d_{i,n}^{B}$ is the distance from the back of the vehicle n in type i to the back of the vehicle that has just passed it on the target lane; $d_{i,n}^{opp}$ is the gap from the front of the vehicle to the front of the closest one on the other lane; gap_i^{B} and gap_i^{opp} account for the minimal back and front gaps for vehicle type i ensuring the safety of lane changing behavior, respectively (Fig. 3).

The second criterion for lane changing is that the operational velocity of the front vehicle is slower than the vehicle's expected velocity in the next time step, and the driver expects a better driving condition on the other lane. Such situation becomes more complicated when there are multiple vehicles ahead (e.g. two bicycles in front of the vehicle n in Fig. 3). To account for such situation, we deem that the speed criterion is satisfied if the minimal expected speed of the front vehicles is lower than the expected velocity of current vehicle, given by

$$\min\{v_{i,\max}, v_{i,n}+1\} > \min\{v_{j,n+1}+1,$$
$$v_{j,\max}, v_{k,n+2}+1, v_{k,\max}\} \quad \forall d_{n,n+1} = d_{n,n+2}$$

where $dn,n+1$ and $dn,n+2$ represent the distances from vehicles $n+1$ and $n+2$ to vehicle n, respectively; $vi,\max$ is the maximal velocity set for the vehicle type i.

Since the proposed model allows multiple vehicles operating in parallel, it is possible that a vehicle can overtake another one without lane changing. For example in Fig. 3, the distance from the vehicle n to its front vehicle (i.e. di,n) is equal to 2 instead of 0 because there is sufficient space in the cell occupied by the vehicle $n+$ ' 1. The same criterion also applies for the calculation of $d_{i,n}^{opp}$, and $d_{i,n}^{B}$. Furthermore, we adopt the concept of positional discii line which implies that in a mixed traffic flow environment, bicycles keep to the side of the road nearest to the curb, while autos accommodate space for any present bicycles by staying as far away from the curb as possible without crossing the median line.

Considering safety, we also assume that when the space is sufficient $(d_{i,n}^{op} \geq 0)$, the vehicle making a change to the opposing lane will immediately go back to the farthest location on the original lane with respect to its velocity. If there is no space allowing vehicles changing to the opposing lane to move back ($d_{i,n}^{op} < 0$, e. g. the vehicle n in Fig. 3), the vehicle will keep driving on the opposing lane until there is a gap.

Although most CA studies have modeled single-lane (Nagel and Schreckenberg, 1992, Fukui and Ishibashi, 1996 and Vasic and Ruskin, 2012) or multiplelanes (Nishinari and Takahashi, 1998, Jiang et al., 2004 and Meng et al., 2007) operations in one direction and have incorporated speed randomization factors to reflect the changes in speed due to uncertainties, it might not be sufficient to represent the impact of opposing flows under a bi-directional environment. As drivers may be more cautious and will decrease the speed with a higher probability to avoid the conflict with the opposing flows, we propose to set the randomization factor for type i vehicles that are experiencing or have just experienced a crossing on the opposing lane to be p'_i. For those vehicles that are not facing crossings, their randomization factor is set to be p_i. However, this rule is not imposed on vehicles driving on the opposing lane when they are overtaking because of no conflict between the flows on the original lane.

In summary, the evolvement rules in the proposed model are given by

Step 1:
Acceleration

$$v_{i,n} = \min\left\{v_{i,n} + 1, v_{i,\max}\right\}$$
$$\forall v_{i,n} \in \left\{0,1,2,\cdots,v_{i,\max}\right\}$$

For the vehicle n in the opposing lane

$$v_{i,n} = \min\left\{v_{i,n} + 1, v_{i,\max}\right\}$$
$$\forall v_{i,n} \in \left\{0,1,2,\cdots,v_{i,\max}\right\}$$

Step 2:
Deceleration

$$v_n = \min\{v_n d_n\} \, \forall \, d_n \in \{0\,1\,2\cdots N\}$$

For the vehicle n in the opposing lane

$$v_{i,n}^{\text{opp}} = \min\{v_{i,n}^{\text{opp}},\, d_n\}$$

Step 3:
Randomization

$$v_{i,n} = \max\left\{v_{i,n} - 1, 0\right\}$$

with the probability p'_i for the vehicle n experiencing or experienced a crossing on the opposing lane, otherwise with a probability P_i.
For the vehicle n in the opposing lane

$$v_{i,n}^{\text{opp}} = \max\left\{v_{i,n}^{\text{opp}} - 1,\, 0\right\}$$

with a probability p_i.

Step 4:
Lane changing

For the vehicle n in the original lane, if

$$\min\left\{v_{n,\max}, v_{i,n} + 1\right\} > \min\{v_{j,n+1} + 1, v_{j\max},$$
$$v_{k,n+2} + 1, v_{k,\max}, \cdots\} \quad \forall d_{n,n+1} = d_{n,n+2} = \cdots$$
$$d_{i,n}^{B} \geq gap_i^{B}$$
$$d_{i,n}^{\text{opp}} \geq gap_i^{\text{opp}}$$

then, the vehicle changes the lane with a probability p_i^{opp} and $vi,n = vi,n^{\text{opp}}$.
For the vehicles on the opposing lane

$$v_{i,n} = v_{i,n}^{\text{opp}} = \min\{\overline{d_{i,n}^{\text{opp}}}, v_{i,n}^{\text{opp}}\} \quad \forall\, d_{i,n}^{\text{opp}} \geq 0$$

Step 5:
Vehicle movement

For the vehicle n that operates or comes back to the original lane

$$x_n = x_n + v_{i,n}$$

For the vehicle n on the opposing lane

$$x_n = x_n + v_{i,n}^{\text{opp}}$$

where $v_{i,n}^{\text{opp}}$ indicates that the velocity of the vehicle n of type i on the opposing lane.

MODEL CALIBRATION AND VALIDATION

This section details the calibration of the esential model parameters and sets the following criteria for data collection site selection:

1) Low impact of pedestrians;
2) Low impact of signalized intersections: intersection signals are not explicitly modeled in our study, thus the data collection site should be far away from signals;
3) Low impact of roadway parlcing.

According to the above requirements, we performed data collection during the peak periods (7 am-9 am and 5 pm-7 pm) on August 27, 2009 at a twolane bi-directional road segment of Chengxian Road, Nanjing City, China. As illustrated in Fig. 4, we used cameras to record the traffic data including the density, velocity, volume and the proportion of large vehicles, cars, and bicycles every 15 s. The width of each lane is 3 m ($N = 3 \times 2 = 6$ m), and $L_{AB} = 15.5$ m.

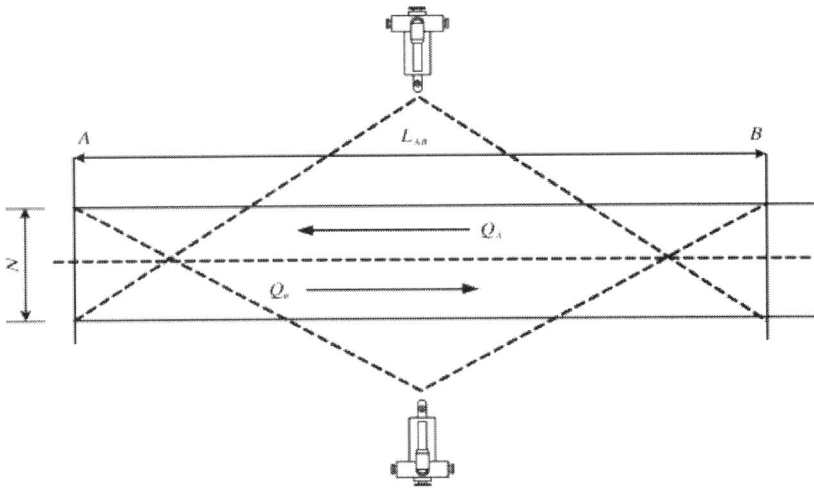

Figure 4. Illustration of data collection method.

NUMERICAL STUDY

Model Installment

In simulation, a system of 2500 cells is considered under the periodic boundary condition. According to Section **2**, the size of each cell is set to be 2.75 m by 3.00 m, so the system is equivalent to a road of 6. 9 km long. When we start to perform a numerical simulation, the types of vehicle with a given density are initially distributed randomly on the road. After a transition time period $t_0 = 1000$ time steps (each time step equals 1 s in the real world), we start to record the time-averaged velocity of traffic flow and the total volume in every period of T (1000 time steps). Finally, we obtain the average velocity and total volume in a run, i. e. the density of total traffic flow is

$$\rho = \sum_i \rho_i$$

The density for each type vehicle is

$$\rho_i = CN_i size_i / 2L$$

The total average velocity is

$$\bar{V} = \sum_i \sum_n v_{i,n} b / N_i$$

The total volume is

$$q = \sum_i \sum_n \rho_i b v_{i,n} / N_i$$

where p_i is density of vehicle type i; N_i and L denote the total number of vehicle in type I and length of road segment, respectively; $size_i$ is the size of vehicle type i (for bicycle, $size_i$ is $1 \times 1/3 = 1/3$, for big vehicle, $4 \times 1 = 4$, and for car, $2 \times 2/3 = 4/3$); the parameter C is used to convert the non-dimensional density into a dimensional form and equals to 363. 4 ($C = 1000/2. 75$); b is used to change the velocity measured by cells into meters; q is total volume with the consideration of car equivalence principle for mixed traffic flow in China (Ren et al. 2003).

Together with the observation and calibration process using the field data, the key parameters, including maximal velocity, randomization probabilities and lane changing probability for each type of vehicle, are given in Tab. 1. Accordingly, the parameter b used to change the velocity measured by cells into meters is set to be 4. In Tab. 2, a volume comparison between the field data and simulation result is given indicating an average error of 4.7 %, where the M/S represents the ratio of big vehicle to car.

Table1. Essential parameters.

Vehicle type	$vi_{,max}$	p_i	pi^{opp}	$p'i$	$gapi^B$	$gapi^{opp}$
Car	12 m/s (4 cells)	0.18	0.2	0.26	6 m (3 cells)	21 m (7 cells)
Bicycle	6 m/s (2 cells)	0.13	0.4	0.2	3 m (1 cells)	12 m (4 cells)
Big Vehicle	9 m/s (3 cells)	0.18	0.25	0.26	6 m (2 cells)	27 m (9 cells)

Table 2. Volume comparison between simulation result and field data.

Bicycle ratio	M/S ratio	Density	Simulation result(vph)	Field data(vph)	Error(%)
0	3:01	0.4	2722	2688	1.26
0.1	2:01	0.22	1375	1396	1.5
0.2	1:01	0.23	952	912	4.39
0.4	1:02	0.31	1411	1310	7.71
0.6	1:03	0.33	1533	1478	3.72
0	3:01	0.35	2736	2789	1.9
0.1	2:01	0.19	1433	1502	4.59
0.2	1:01	0.45	1750	1688	3.67
0.4	1:02	0.32	1928	1875	2.83
0.6	1:03	0.26	1514	1526	0.79
0	3:01	0.28	1922	1847	4.06
0.1	2:01	0.12	1222	1310	6.72
0.2	1:01	0.22	987	957	3.13
0.4	1:02	0.32	1497	1526	1.9
0.6	1:03	0.18	1475	1447	1.94
0	3:01	0.2	1314	1322	0.61
0.1	2:01	0.33	1369	1392	1.65
0.2	1:01	0.17	956	914	4.6
0.4	1:02	0.22	1272	1181	7.71
0.6	1:03	0.15	1198	1101	8.81

Numerical Analysis

In order to capture the consecutive change in operational condition of mixed traffic flow due to the variation of vehicle composition, we increase the proportion of bicycles based on a given ratio of the big vehicles (buses/trucks) to cars, and then smoothly add more vehicles until all cells are occupied.

Volume-density Analysis

Fig. 5, Fig. 6 and Fig. 7 illustrate the volume-density analysis results: 1) The maximal volume increases firstly and decreases latterly with the increase of density. 2) With the increase in the proportion of the big vehicle the total volume decreases. As big vehicles require larger spaces for operations while asociated with a relatively low speed, all vehicles therefore have a lower probability to conduct overtakes but follow their front ones with the consideration of safety. 3) Bicycles have a great impact to the traffic condition in terms of the total volume. The total volume drops dramatically by 30% after we add 10% bicycle into the flow which is only consisted by autos (big vehicles and cas). As bicycles have more flexibility with smaller sizes and higher probabilities in lane changing when the safety criteria are met, other vehicles therefore are easily to be inferred and forced to follow them with a relatively low speed. 4) The maximal total volume stars to recover when the proportion of bicycle is higher than 10%. Though the bicycles have a low velocity, their advantages lying on the low requirement for the space can compensate such deficiency and increase the traffic volume by adding more bicycles into the network. 5) The phenomenon of double-summit in terms of the total volume appears when the proporion of bicycle is beyond 60%. One reasonable explanation is that for the first summit, with the increase of the density (lower than 0.2), vehicles with higher velocity may change their lanes to avoid the interference caused by their front vehicles with a relative low speed. While for the consideration of safety requirement, the total volume drops along with decrease in opporunities of lane changing when the density is higher than 0. 2. However, when the density approximately reaches 0.45, the whole traffic may operate with a low speed and bicycles therefore can find more chances to change their lanes and eventually lead to the increase in total volume.

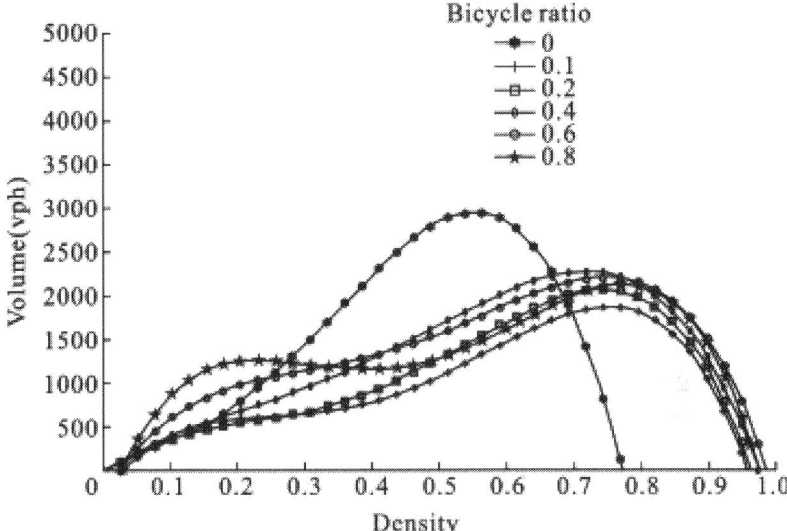

Figure 5. Volume-density curves when the ratio of big vehicles to cars is 3:1.

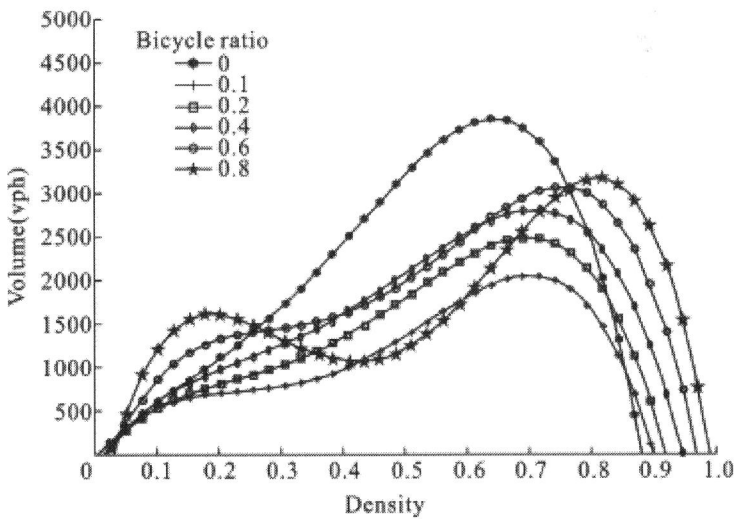

Figure 6. Volume-density curves when the ratio of big vehicles to cars is 1:1.

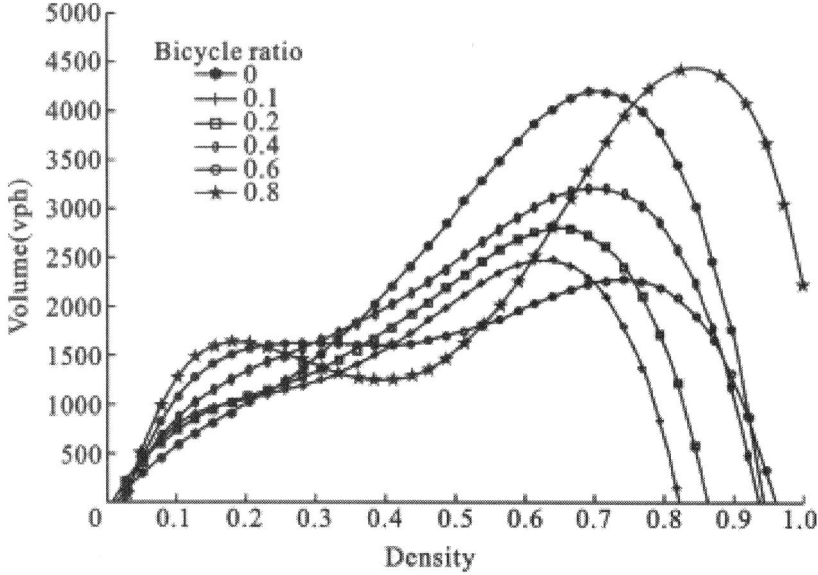

Figure 7. Volume-density curves when the ratio of big vehicles to cars is 1:3.

Velocity-density Analysis

As shown in Fig.8, Fig.9 and Fig. 10, we illustrate the relation between velocity and density with different compositions of vehicle types. The following findings can be reached; 1) Bicycles have a great impact on the average velocity. Compared with the velocity of traffic flow only consisted by big vehicles and cars, the operational speed drop 60% approximately when the bicycle is added. 2) The variation of velocity changes in a small range when the proportion of bicycle is higher than 10%. Because most autos (big vehicles and cars) are following their front vehicles with a slow speed after the incorporation of bicycles, the space requirement for bicycles moving with their maximal speed is relatively easy to be satisfied. The average speed of the flow is therefore concentrated around the maximal velocity of bicycles. 3) As the cars have higher maximal velocities, the increase in the proportion of cars will lead to the increase in total average speed though it is not signifiant when the bicycles are taken into account.

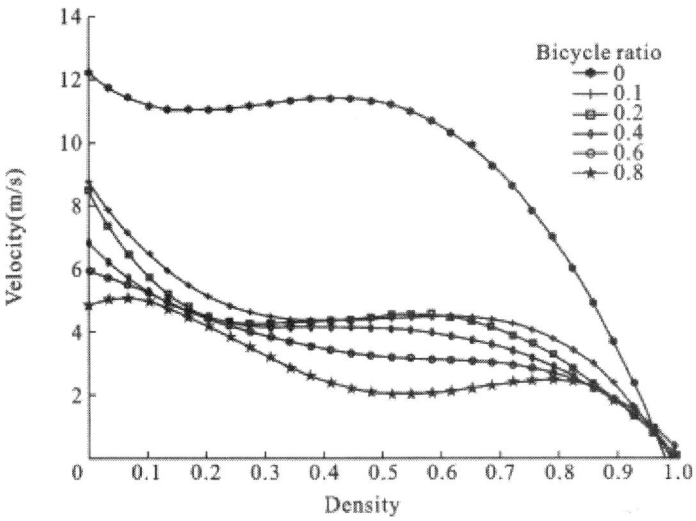

Figure 8. Speed-density curves when the ratio of big vehicles to cars is 3:1.

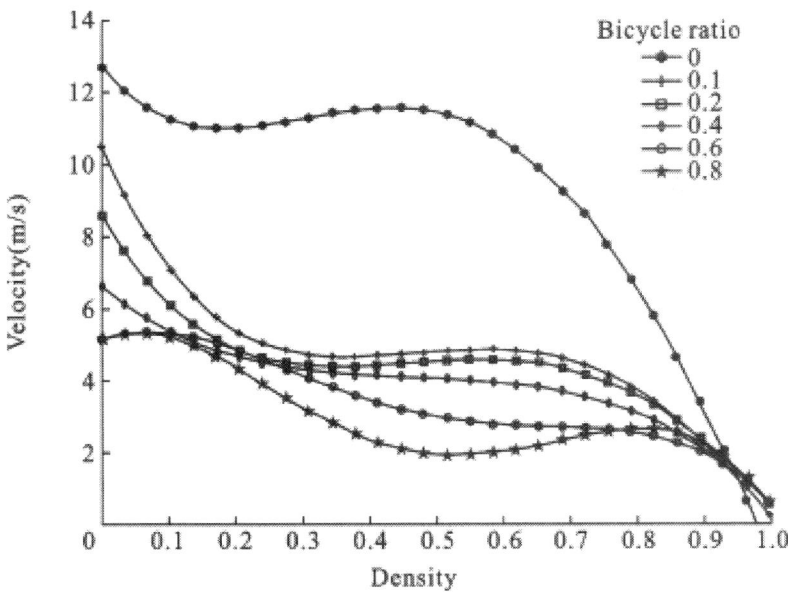

Figure 9. Speed-density curves when the ratio of big vehicles to cars is 1:1.

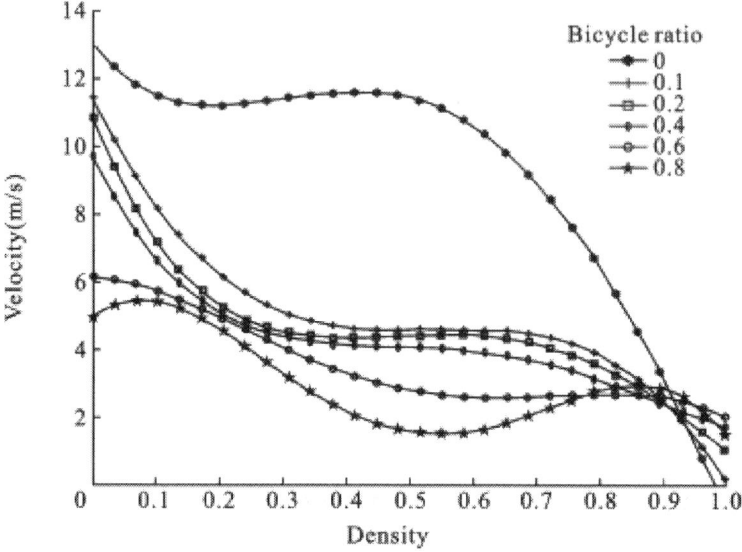

Figure 10. Speed-density curves when the ratio of big vehicles to cars is 1:3.

CONCLUSIONS

In summary, in this paper, we build a CA model to investigate the operational characteristics of mixed traffic flow at a two lane bi-directional road segment. To capture more reality, we categorize vehicles into different groups including big vehicles (buses/trucks), cars, and bicycles, and take into account their physical and mechanic differences. Essential parameters with regards to different vehicle types are calibrated by using the field data collected at a two-lane bi-directional road segment in Nanjing City, China. We have detailed the relations of volume-density and velocity-density under the various compositions of traffic flow which leads to some key findings including: 1) Bicycles impact the traffic conditions in terms of volume and velocity to a large extent that is the average speed and maximal volume drop 60% and 30% respectively after the incorporation of 10% bicycles into the mixed traffic flow. 2) The volume stars to recover when the proporion of bicycles is higher than 10%. 3) The phenomenon of double summits of the traffic volume appears when the proportion of bicycles is higher than 60%. 4) The average speed of total traffic flow is highly concentrated after the incorporation of bicycles.

In the next step, this model will be extended to incorporate signals at intersection, and queuing behavior and stop-and-go conditions will be explicitly modeled. Behaviors and impact of pedestrian's will be taken into account to describe more complex traffic flow at both intersections and road segments. Furher, the situation of multiple lanes (more than 2) will be considered as which may induce other lane changing behaviors having a great influence on the traffic conditions (i. e. change to another in the same direction).

REFERENCES

1. Adams, W. F., 1936. Road traffic considered as a random series. Journal of the ICE, 4(1): 121–130.
2. Basso, L. J., Guevara, C. A., Gschwender, A., et al., 2011. Congestion pricing, transit subsidies and dedicated bus lanes: efficient and practical solutions to congestion. Transport Policy, 18 (5): 676–684.
3. Chander, R. E., Herman, R., Montroll, E. W., 1958. Traffic dynamics: studies in car following. Operations Research, 6 (2): 165–184.
4. Cremer, M., Ludwig, J., 1986. A fast simulation model for traffic flow on the basis of boolean operations. Mathematics and Computers in Simulation, 28 (4): 297–303.
5. Foulaadvand, M. , Belbasi, S., 2007. Vehicular traffic flow at a nonsignalized intersection. Journal of Physics A: Mathematical and Theoretical, 40 (29): 8289–8297.
6. Fukui, M. , Ishibashi, Y., 1996. Traffic flow in 1D cellular automaton model including cars moving with high speed. Journal of the Physical Society of Japan, 65(6): 1868–1870.
7. Greenshields, B. D., Shapiro, D., Erickson, E. L., 1947. Traffic performance at urban street intersections. Bureau of Highway Traffic, Yale University, New Haven.
8. Gundaliya, P., Mathew, T., Dhingra, S., 2008. Heterogeneous traffic flow modelling for an arterial using grid based approach. Journal of Advanced Transportation, 42 (4): 467–491.
9. Herman, R., Montroll, E. W., Potts, R. B., et al., 1959. Traffic dynamics: analysis of stability in car following. Operations Research, 7(1), 86–106.
10. Jiang, R., Jia, B., Wu, Q., 2004. Stochastic multi-value cellular automata models for bicycle flow. Journal of Physics A: Mathematical and General, 37(6)' 2063–2072.

11. Khan, S. I., Maini, P., 1999. Modelling heterogeneous traffic flow. Transportation Research Record, 1678: 234–241.
12. Kin=, J. P., 1993. Application of the theoty of probability tD problems of highway traffic. Traffic Engin=ing and Control, 5, 284–287.
13. Ughthill, M. J., Whitham, G. B., 1955. On kinematic waves ll: a theory of traffic flow on long crowded roads. Proceedings of the Royal Society of London. Series A, Mathematical and Physical Sciences, 229(1178), 317–345.
14. Mallikarjuna, C., Rao, K., 2009. Cellular automata model for heterogeneous traffic. Journal of Advanced Transportation, 43 (3): 321– 345.
15. Meng, J. P., Dai, S. Q., Dong, L. Y., et al., 2007. Cellular automaton model for mixed traffic flow with motorcycles. Physica A: Statistical Mechanics and its Applications, 380(1/2): 470480.
16. Nagel, K., Schreckenberg, M., 1992. A cellular automaton model for freeway traffic. Journal de Physique Aiclrives, 2 (12) , 2221–2229.
17. Nishinari, K., Takahashi, D., 1998. Analytical properties of ultradiscrete Burgers equation and rule-184 cellular automaton. Journal of Physics A, Mathematical and General, 31 (24), 5439–5450.
18. Payne, H. J., 1971. Models of freeway traffic and control. Mathematical Models of Public Systems, 1 (1), 51–61.
19. Payne, H. J., 1979. FREFLO: a macroscopic simulation model of freeway traffic. Transportation Research Record, 722: 68–77.
20. Ren, F. T., Liu, X. M., Rang, J., 2003. Traffic engineering science. China Communications Press, Beijing.
21. Ruskin, H. J., Wang, R., 2007. Lecture notes in computer science, 2329' 381.
22. Vasic, J., Ruskin, H. J., 2012. Cellular automata simulation of traffic including cars and bicycles. Physica A: Statistical Mechanics and its Applications, 391(8), 2720–2729.
23. Xie, D. F., Gao, Z. Y., Zhao, X. M., et al., 2009. Characteristics of mixed traffic flow with non-motorized vehicles and motorized vehicles at an unsignalized intersection. Physica A: Statistical Mechanics and its Applications, 388(10): 2041–2050.
24. Zhao, X. M., Gao, Z. Y., 2005. Controlling traffic jams by a feedback signal. The European Physical Journal B, Condensed Matte< and Complex Systems, 43(4) , 565–572.

CITATION

Zhenke Luo, Yue Liu, Chen Guo, Operational characteristics of mixed traffic flow under bi-directional environment using cellular automaton, Journal of Traffic and Transportation Engineering (English Edition), Volume 1, Issue 6, December 2014, Pages 383-392, ISSN 2095-7564, http://dx.doi.org/10.1016/S2095-7564(15)30288-9.

CHAPTER 2

Traffic Flow Consideration in Design of Freight Distribution System

Sutanto Soehodho, Nahry

Dept. of Civil Engineering, Faculty of Engineering, University of Indonesia, Kampus UI Depok 16424, Indonesia

ABSTRACT

This study is part of a series of research projects on a distribution system we developed to deal with cases in a state-owned company. It concerns the design of the Public Service Obligation State-owned Company (PSO-SOC) distribution system. The intrinsic features of PSO-SOC are distributing strategic commodities and having subsidies within the cost function. Hence their distribution flow has to be secured under consideration of moving the commodities within road networks that have traffic flow dependency. This paper focuses on the solution of the proposed model which represents traffic flow dependency within a freight distribution network.

The mathematical formulation takes the form of a Minimum Cost Multicommodity Flow (MCMF) problem. Traffic flow dependency is incorporated into the model by introducing a coefficient of speed, which is derived from the traffic assignment of ordinary traffic associated with the transportation of the type of freight under consideration The solution of the proposed model is formulated by Network Representation (NR), in which all of the components of the mathematical model are represented in the form of dummy links and nodes added to the original (physical) network. It is to be noted then, that the traffic flow on each road or link is represented by a link performance function (LPF), depicting traffic flow dependent travel time and consequent cost. The MCMF problem of NR is further solved by a Primal–Dual Algorithm.

Finally, an illustrative example is exercised to show how the proposed step-wise solution works.

INTRODUCTION

Research on designing freight distribution systems has been done for many years. Most deal with a private company whose primary concern is merely profit maximization. In Indonesia, there exists a subsidized service by State-Owned Company (SOC) called Public Service Obligation (PSO)-SOC. PSO-SOC has the obligation to serve the entire demand on public commodities or services. Its working orientation is not for profit, but for security of supply. The PSO-SOC is still permitted to conduct its own programs beyond its main task, but it is undertaken within government control and limitation. The PSO-SOC also bears strategic commodities in distribution, so availability of commodities at the right time and place is important. Most of those commodities are being transported through surface land transport with mixed-traffic, so travel time and cost are very much dependent on the flow of vehicles within the road network.

The identification of the distribution system of one of Indonesia's PSO-SOCs, which deals with the production and distribution of public commodities, provides insight to some important issues as follows [1]:

1. The company under consideration is a group of companies that consists of 1 (one) holding company and 5 (five) affiliated companies. Each of the companies (included the holding company itself) carries out the operation of its own plant and its distribution process independently. Those companies are managed separately and there is no regulation that integrates those companies in their logistical process.
2. Unit production costs are not uniform among the plants. This is due to the different prices of raw material and the variability in the operation performance of the production processes.
3. Product differentiation exists. This implies that products are not differentiated merely by type (material) of product, but also by the type of user. There are two types of user, public (subsidized) and commercial. Both of them are different in terms of selling price and

demand satisfaction. Subsidized prices are determined by the government, while the commercial ones are set by the company. Naturally, the commercial prices are higher than the subsidized ones. Moreover, subsidized demands have the privilege of being fully satisfied regardless of the amount of profit that the company may receive from them.

Previous research on freight distribution systems was concerned mostly with private companies [2], [3] and [4]. Most of the research on the distribution of public needs is related to public services (such as schools, police stations, hospitals, etc.) rather than public goods. Savas [5] focused his research on the equity in providing public services, while Ross [6] proposed a model with multicriteria to select sites for public facilities. Regarding the variables included, most of the research focused on transportation costs, while some dealt with various other variables, such as production cost, fixed cost of facility, inventory holding cost, and anything relevant to the special problem they faced[2], [7], [8] and [9].

Moreover, most of the research on freight distribution systems hardly consider the effect of ordinary traffic movement on the system. The FHWA reports [10] that the private and public sectors of the freight industry bear an operational cost of congestion which is as much as $25–200 an hour, depending on product type and other factors. The report also estimates that unexpected delays can increase the cost of transporting goods by 50–250%. Jones [11] explains that for the freight industry and trucking companies, congestion on the transportation network diminishes productivity and increases the overall cost of transportation services significantly. Other effects of traffic delays are a higher cost for fleet operations, decreased fleet and vehicle utilization, decreased fuel efficiency, increased emissions due to idling, and decreased hours of "productive" service for drivers.

These intrinsic characteristics clarify the need to enhance the earlier distribution models for the purpose of taking into account the special role of a state-owned company and considering the movement of the commodities within road networks that have traffic flow dependency.

MODEL FORMULATION

In order to take into account the characteristics of the PSO-SOC as stated in Section 1, we propose that the distribution system of the affiliated companies should be integrated into one system which is coordinated by the holding company. Furthermore, we propose a distribution model which deals with production cost, transportation cost, warehouse cost, as well as revenue as its variables [1]. These variables are considered for the following reasons. When the total plant capacity is more than the total demand, the holding company has to designate production allocation to each plant, and in such a situation, one or more plants must be under capacity (not operated in full capacity). When the holding company only considers the distribution cost, it may lose efficiency on logistics as a whole, particularly if the cost of production is not in accordance with the efficiency at the distribution cost level.

In the case of over demand, all of the plants must be fully operated. In such case, the holding company must decide which commercial demands should be satisfied in order to attain the maximum profit of the company. Since the selling price of commercial products varies significantly, the selling price becomes a very important variable to be included in this optimization.

Regarding ordinary traffic considerations, we propose that a coefficient of speed be included in the model. The coefficients of speed are derived from ordinary traffic assignments which produce user-equilibrium link travel time and the associated speeds. It is actually quite valuable to consider the effect of ordinary traffic on the movement of freight vehicles, since it will lead to more informed and efficient decisions for freight allocation. Indeed, the decision to assign some amount of products to a set of plants and distribute them to a set of end consumers through certain paths is crucial to maintaining low transportation costs since these costs are highly affected by traffic performance on any given path.

The PSO-SOC under consideration takes into account only transportation costs and it does not consider the ordinary traffic flow effect in its product allocation process. Besides, the process is carried out by each plant independently (each plant is operated exclusively by one affiliated company) and there is no integration with the holding company.

In order to be more precise in making decisions on product allocation, we propose a model which deals with the cost of producing, transporting, and handling the commodities [1], as well as revenue, and includes the effect of ordinary traffic on the product allocation decision. Product allocation is also proposed to be optimized by integrating logistic subsystems of all plants.

In order to cope with the problem of a distribution system characterized mainly by product/demand differentiation, integrated systems, as well as traffic flow dependency, we propose a mathematical model which is applicable to the following distribution network, as depicted in Fig. 1. It consists of a set of plants, consolidation centers and retailers.

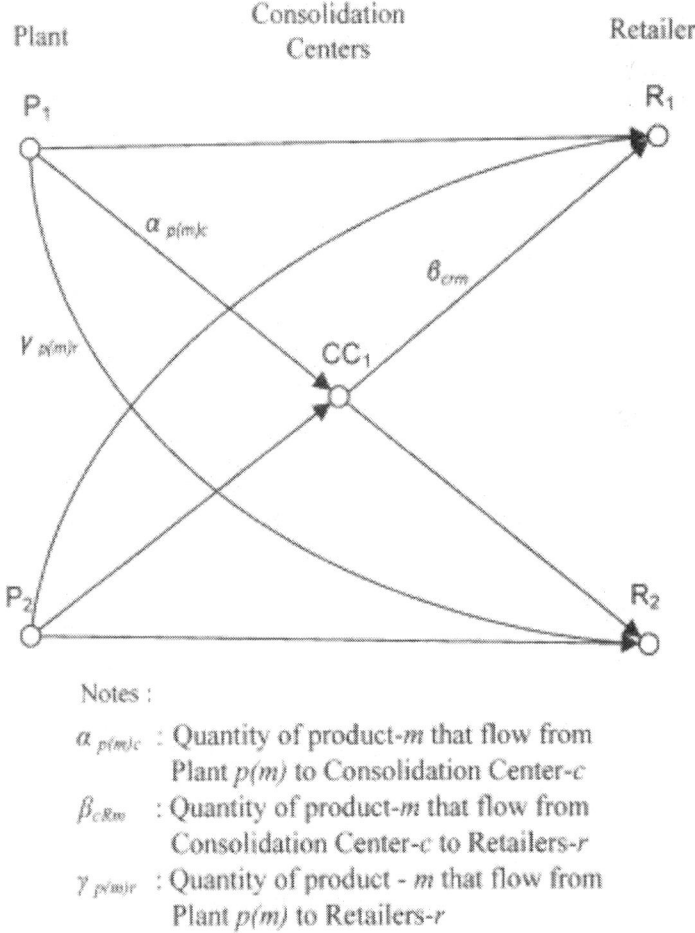

Notes :

$\alpha_{p(m)c}$: Quantity of product-m that flow from Plant $p(m)$ to Consolidation Center-c

β_{cRm} : Quantity of product-m that flow from Consolidation Center-c to Retailers-r

$\gamma_{p(m)r}$: Quantity of product - m that flow from Plant $p(m)$ to Retailers-r

Figure 1. An example of a physical distribution network.

The proposed mathematical model is as follows:

$$
\begin{aligned}
\min Z(\alpha_{p(m)c}, \beta_{crm}, \gamma_{p(m)r}) = & \sum_{p\in P}\sum_{c\in C}\sum_{m\in M}\mu_{pc}\cdot d_{pc}\cdot u_{pcm}\cdot \alpha_{p(m)c} \\
& + \sum_{c\in C}\sum_{r\in R}\sum_{m\in M}\mu_{cr}\cdot d_{cr}\cdot v_{crm}\cdot \beta_{crm} \\
& + \sum_{p\in P}\sum_{r\in R}\sum_{m\in M}\mu_{pr}\cdot d_{pr}\cdot z_{prm}\cdot \gamma_{p(m)r} \\
& + \sum_{p\in P}\sum_{c\in C}\sum_{m\in M}w_{cm}\cdot \alpha_{p(m)c}\sum_{p\in P}\sum_{m\in M} \\
& \times \left(\sum_{c\in C}\alpha_{p(m)c} + \sum_{r\in R}\gamma_{p(m)r}\right)\cdot \eta_{p(m)} \\
& - \sum_{r\in R}\sum_{m\in M}\left(\sum_{c\in C}\beta_{crm} + \sum_{p\in P}\gamma_{p(m)r}\right)\cdot \rho_{rm}
\end{aligned}
$$

(1)

Turn MathJax on

$$
\mu_{ij} = 1 - \frac{v_{UEij} - v_D}{v_D}
$$

(2)

Turn MathJax on

subject to

$$
\sum_{p\in P}\alpha_{p(m)c} = \sum_{r\in R}\beta_{crm}, \qquad \forall c\in C, \forall m\in M
$$

(3)

Turn MathJax on

$$
\sum_{c\in C}\beta_{crm} + \sum_{p\in P}\gamma_{p(m)r} = \lambda_{rm}, \qquad \forall r\in R, \forall m\in M_s
$$

(4)

Turn MathJax on

$$\sum_{c \in C} \beta_{crm} + \sum_{p \in P} \gamma_{p(m)r} \leq \lambda_{rm}, \qquad\qquad \forall r \in R, \forall m \in M_c \tag{5}$$

Turn MathJax on

$$\sum_{c \in C} \sum_{m \in M} \alpha_{p(m)c} + \sum_{r \in R} \sum_{m \in M} \gamma_{p(m)r} \leq Cp_{p(m)} \quad \forall p \in P, \forall m \in M \tag{6}$$

Turn MathJax on

$$\alpha_{p(m)c} \geq 0, \qquad\qquad \forall p \in P, \forall c \in C, \forall m \in M \tag{7}$$

Turn MathJax on

$$\beta_{crm} \geq 0, \qquad\qquad \forall c \in C, \forall r \in R, \forall m \in M \tag{8}$$

Turn MathJax on

$$\gamma_{p(m)r} \geq 0, \qquad\qquad \forall p \in P, \forall r \in R, \forall m \in M \tag{9}$$

Turn MathJax on

Subscripts

p:

indicate the plants

c:

indicate the consolidation centers

r:

indicate the retailers

m:

indicate the products

$p(m)$:

indicate the plant $p \in P$ that produces product-m

Sets

P:

set of plants

C:

set of consolidation centers

R:

set of retailers

M:

set of products

$Ms \in M$:

set of subsidy (public) products

$Mc \in M$:

set of commercial products

Decision variables:

$\alpha p_{(m)c}$

is quantity of product-m that flow from plant $p(m)$ to consolidation center-c

βcrm

is quantity of product-m that flow from consolidation center-c to retailer-r

$\gamma p_{(m)r}$

is quantity of product-m that flow from plant $p(m)$ to retailer-r

Input parameters

upcm:

per-mile cost to ship a unit of product-m from plant-p to consolidation center-c

vcrm:

per-mile cost to ship a unit of product-m from consolidation center-c to retailer-r

zprm:

per-mile cost to ship a unit of product-m from plant $p(m)$ to retailer-r

d_{ij}:

length of distance of link $i–j$

μ_{ij}:

coefficient of speed of link $i–j$

$vUEij$:

user-equilibrium speed of link $i–j$

wcm:

unit warehouse cost to handle product-m in consolidation center-c)

$\eta_{p(m)}$:

unit cost for producing product-m in plant- p

prm:

selling price of product-m at retailer-r

$Cpp_{(m)}$:

capacity of plant-p to produce product-m

demand of product-m in retailer-r

design speed of link $i–j$

Eq. (1) denotes the objective function of the proposed model. It actually maximizes the profit, in which profit is represented by revenue minus cost.

Similarly, this objective function can be replaced by a minimization of cost function, which is represented by cost minus revenue.

The first three terms of Eq. (1) represent transportation cost, in which each term includes a coefficient of speed. Each coefficient of speed is exclusive for a certain link. It is derived from the user-equilibrium speed of an ordinary traffic assignment and design speed, as formulated in Eq. (2). The use of coefficients in Eq. (1) is meant to indirectly make corrections to the design unit cost through a correction of the design speed due to the dynamics of traffic conditions in real life situations. Since traffic conditions are very dynamic in nature, the time windows considered should be carefully selected. Moreover, this coefficient indicates that the smaller the speed (the more congested the road) the longer the "distance" that the freight vehicle should travel. The fourth term of Eq. (1) is related to warehouse cost. The fifth term represents production cost and the last term concerns revenue. Obviously, due to the opposite characteristics of cost and revenue, we must put a minus sign before revenue. Eq. (3) denotes that the total inflow minus the total outflow in consolidation centers is set at zero since those nodes are intermediate nodes.

Eqs. (4) and (5) are related to demand satisfaction for subsidized products and commercial ones, respectively. Subsidized products must be entirely fulfilled, while the commercial ones could be satisfied later, in the case of excess plant capacity. Eq. (6)implies that the total amount of production of any product by each plant should not be more than its capacity. Eqs. (7), (8) and (9) ensure non negativity of flow constraints.

MODEL SOLUTION

Since the MCMF problem is highly related to the network structure, we utilize Network Representation (NR) to represent and solve the proposed mathematical model. Network Representation (NR) is a technique used to solve a problem by representing a mathematical model as a network flow-based formulation [12]. It is characterized by the use of diagrams that have emerged, by progressive expansion, from those used traditionally in network flow and graph theory. Network Representation is developed by adding some dummy links and nodes into the original (physical) network, in which the function of those dummy links is designated to represent production cost, transportation cost, and warehouse cost, as well as revenue.

One particular issue regarding our MCMF problem is the involvement of multi-commodities in the distribution system. This issue is solved by introducing a Sub-Network Representation devoted to a certain product. We name such sub-NRs as Product Sub-Network Representation (P-SNR). Each P-SNR is exclusively devoted to a certain product, and consists of nodes and links used by that certain product, although it is possible that some of the links of a P-SNR are used for common products.

In order to cope with the situation of an imbalance between total supply and total demand, we developed an Excess Supply/Demand Sub-Network Representation. Its function is to balance the total supply and total demand endogenously in the optimization process, by decreasing the demand or supply so that both of them finally are in balance.

An example of a general NR is shown in Fig. 2. It consists of 2 plants, 1 consolidation center and 2 retailers, and it deals with 2 types of products and 2 types of demand (subsidized and commercial demand). The P-SNRs of such an NR are shown in Fig. 3a and b.

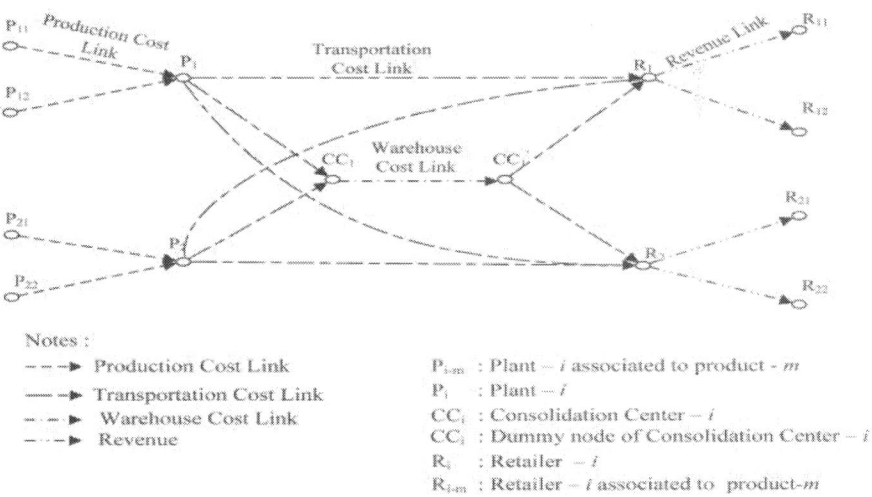

Notes :

— – ▶ Production Cost Link P_{i-m} : Plant – i associated to product - m
——— ▶ Transportation Cost Link P_i : Plant – i
— ·· ▶ Warehouse Cost Link CC_i : Consolidation Center – i
— ·· ▶ Revenue CC_j : Dummy node of Consolidation Center – i
                                    R_i : Retailer – i
                                    R_{i-m} : Retailer – i associated to product-m

Figure 2. An example of a Network Representation.

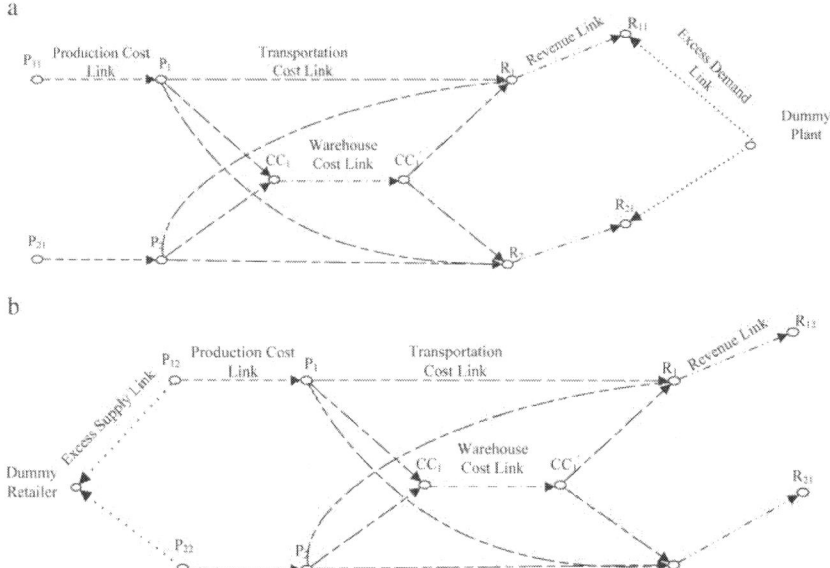

Figure 3. a. Product Sub-Network Representation of product-1 (in Excess Demand case). b. Product Sub-Network Representation of product-2 (in Excess Supply case).

Links between node Pi_m and Pi are designed as production cost links. Each link represents the cost to produce product-m in plant-i. Each of those links is also characterized as a product-exclusive link; that is each link is devoted to a certain product. Hence, in Fig. 3a, which shows the P-SNR of product-1, there are only nodes and links that relate to product-1 (link P_{11}–P_1, P_{21}–P_2).

Links between Pi–CCi, Pi–Ri and CCi'–Ri are designed as transportation cost links, and they represent the transportation cost between two distribution facilities. It is assumed that the unit cost to transport any type of product in a certain link is similar. Coefficients of speed are employed in each transportation cost link. The link between CCi–CCi' represents the cost of using warehouse-i. Every one unit of flow that comes to the consolidation center is charged by one unit of warehouse cost. Links between node Ri and Ri–m are designed as revenue links. Those links represent revenue from selling product-m to retailer-r. Each revenue link is also designed as a product-exclusive link. Unit cost associated to revenue link is denoted by selling price. All of the links of the NR are designed as un-capacitated links; hence there will be as many flows as possible passing through the links.

In a real situation, the total supply is not always in balance with the total demand. In some cases, the total supply is higher than the total demand. In such excess supply cases, the company should be selective in making product assignments to each plant. In excess demand cases, in which the total demand is higher than the total supply, the company should be selective in demand satisfaction. In the case of the PSO-SOC, subsidized demand should have priority to be fulfilled, no matter what the profit that the company could attain from it. Regarding both cases, we developed an Excess Supply/Demand Sub-Network Representation in order to accommodate both situations and to make the supply and demand be always in balance during the optimization process. Fig. 3a depicts the NR in the excess demand case, in which we added a Dummy Plant to act as a supply for the excess demand. Fig. 3b depicts the excess supply case, in which we added a Dummy Retailer to act as a receiver for some excess products from nodes of plants.

In order to give priority to the subsidized demand, we designate an extremely high unit cost to the links of the excess Demand Sub-Network that relates to nodes of subsidized (public) demand. This means that such a high "unsatisfied-demand cost" will cause the model to avoid unfulfillment of public demand.

The nodes of the NR are valued as their flow requirement. Node Pi_m is valued by the capacity of plant-i on producing product-m, meanwhile node Ri_m is valued by the demand on product-m in retailer-i. The flow requirements of the intermediate nodes are set as zero.

Having the NR formulation, the problem now is how to determine the optimal assignment of the products to attain the "minimum cost". The minimum cost flow problem is solved by a Primal–Dual Algorithm [13]. We propose a step-wise solution of the MCMF problem as depicted in Fig. 4. It can be explained as follows:

Step 1:
The process is initialized by doing traffic assignments to the ordinary traffic of the physical network of the freight distribution system. Provide the attributes of the ordinary traffic network, including link capacity, link performance function and O/D flow. Set the time window and do traffic assignments to find the user-equilibrium link travel time and the associated speed. Provide the Coefficient of Speed of each link using Eq. (2).

Step 2:
Develop the General Network Representation (NR) of the freight allocation problem. Define the link capacity and the flow requirement as well as the unit cost of the NR by considering the Coefficient of Speed of step 1.

Step 3:
Develop product sub-NR (P-SNR) of all types of products.

Step 4:
Add an Excess Supply/Demand Sub-Network to the P-SNR of step 3; if it is needed, include setting of its link capacity and unit cost.

Step 5:
Solve the MCF problem of the P-SNR by using a Primal–Dual Algorithm.

Step 6:
Find the optimal flow and its associated paths through the P-SNR. Steps 3–6 are repeated until all products are optimized. Since one P-SNR is independent to another P-SNR, steps 3–6 are actually able to be carried out in a simultaneous way.

Step 7:
When all the P-SNRs have been optimized, superimpose the optimal paths of all P-SNRs and find the total optimal flows on each link of the NR. Superimposing is allowable since it is assumed that unit cost to transport and handle any type of product in a certain link is similar, and each link of production cost, as well as revenue links, is characterized by a product-exclusive link.

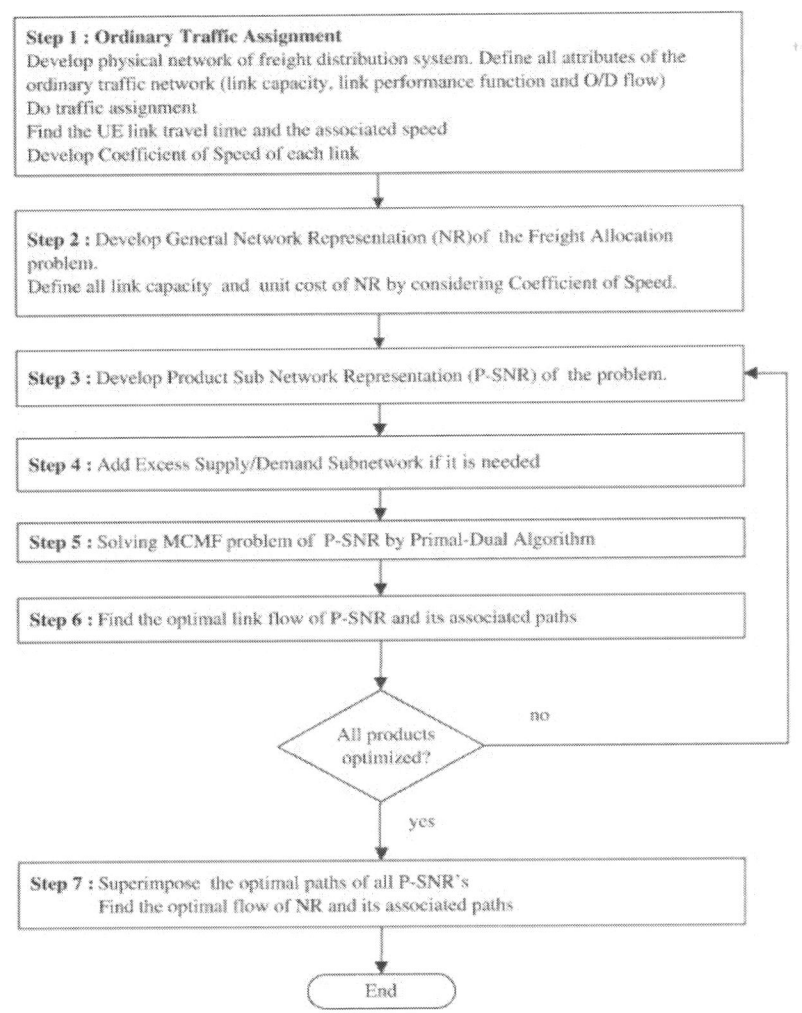

Figure 4. Step-wise of model solution.

ILLUSTRATIVE EXAMPLE

In an attempt to apply the step-wise proposed in Section 4, the ensuing contrived example is discussed. The distribution network of the example consists of 2 plants, 1 consolidation center and 2 retailers. It deals with 2 kinds of products and 2 kinds of demand (subsidy and commercial). Plant capacity and demand on each product are shown in Table 1 and Table 2 respectively.

Table 1. Plant capacity.

Plant	Plant capacity (units)	
	Product-1	Product-2
1	140	100
2	210	660
Total	350	760

Table 2. Demand on certain product.

Retailer	Demand on product (units)			
	1^S	1^C	2^S	2^C
1	50	20	10	30
2	30	10	20	40
Total	110		100	

Notes: s: subsidy; c: commercial.

We exercised this example with 3 cases. In the first case, it is assumed that coefficients of speed of all links are one. In the second and third cases, it is assumed that step 1 of the step-wise model is already done and 2 sets of link coefficients of speed were found. The first set is derived from a traffic assignment which is based on user-equilibrium with an average speed of 58.63 units (case 2) and the second one is based on user-equilibrium with an average speed of 61.2 units (case 3).

The optimal flow of the three traffic assignments is shown in Fig. 5 and the associated product assignment is displayed in Table 3. The values of the objective function, as well as the patterns of product assignment of all cases, are changed as the coefficient of speed changes. It can be concluded that the coefficients of speed are sensitive to the product assignment, as well as the value of the objective function. Moreover, it can be said that ordinary traffic flow is a necessary factor to be considered in designing a freight distribution system model in order to make it more realistic.

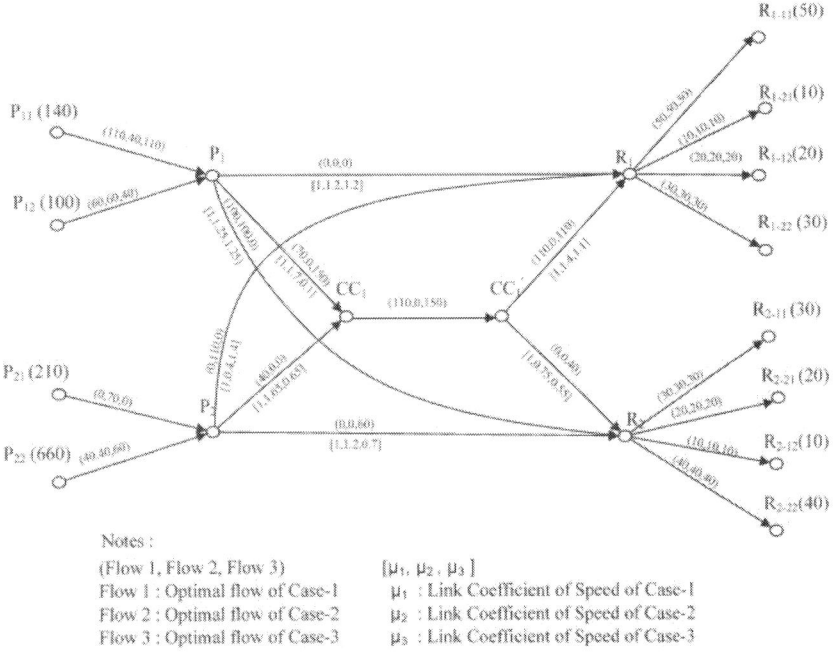

Figure 5. The optimal flows of 3 (three) distribution assignments.

Table 3. The assignment of the products.

Retailer	Product	Supplied by plant		
		Case 1	Case 2	Case 3
		Objective value = − 2670	Objective value = − 2538	Objective value = − 2896
1	1^S	1	2	1
2	1^S	1	1	1
2	1^C	1	1	1
1	1^C	1	2	1
1	2^S	2	2	1
2	2^S	1	1	2
2	2^C	1	1	2
1	2^C	2	2	1
Average speed		60 (Design speed)	58.63	61.2

Notes: s: subsidy; c: commercial.

CONCLUSION

We propose an allocation model which considers traffic flow dependence and the characteristics of the PSO-SOC. Traffic flow, that may increase travel cost through LPF and hence distribution cost, is accounted for within the traffic assignment. By doing so, the flow of commodities within mixed-traffic will implicitly represent cost performance due to the nature of traffic flow situations. The proposed model takes the form of a Minimum Cost Multicommodity Flow problem and the solution is formulated by Network Representation.

Calibration of the model may be required when dealing with real cases in determining the coefficients of speed. The speed coefficient given in the example is a simple one to represent the relation between traffic flow patterns and the pattern of distribution. A more realistic one, which is close to a real condition representation, may be made with an empirical process through model calibration. The three cases given in exercising the model merely denote the different distribution patterns due to different traffic patterns. Those three cases use different values for the objective function, but they all are based on flow equilibrium conditions that can be made to cope with dynamic traffic flows in practical cases.

This research work is essentially intended to give a contribution to the research field of freight distribution within mixed-traffic, as well as to the distribution system of a PSO-SOC within the proposed assumptions.

REFERENCES

1. S. Soehodho, Nahry, Strategic design of distribution system of state-owned companies: preliminary stage of logistics research series and evaluation on model parameters, International Journal on Logistics and Transportation (IJLT) 3 (1) (2009).
2. K.S. Bhutta, F. Huq, G. Frazier, Z. Mohamed, An integrated location, production, distribution and investment model for a multinational corporation, International Journal Production Economics 86 (2003) 201–216.
3. M. Sun, Solving the uncapacitated facility location problem using tabu search, Computers and Operations Research 33 (2006) 2563–2589.
4. L. Dupont, Branch and bound algorithm for a facility location problem with concave site dependent costs, International Journal Production Economics 112 (2008) 245–254.

5. E.S. Savas, On equity in providing public services, Management Science 24 (8) (1978) 800.
6. G.T. Ross, R.M. Soland, A multicriteria approach to the location of public facilities, European Journal of Operational Research 4 (1980) 307–321.
7. S. Yan, D.S. Juang, C.R. Chen, W.S. Lai, Global and local search algorithms for concave cost transshipment problems, Journal of Global Optimization 33 (2005) 123–156.
8. J. Harkness, C. ReVelle, Facility location with increasing production costs, European Journal of Operational Research 145 (2003) 1–13.
9. E.T. Iakovou, An interactive multiobjective model for the strategic maritime transportation of petroleum products: risk analysis and routing, Safety Science 39 (2001) 19–29.
10. K. White, L.R. Grenzeback, Understanding freight bottlenecks, Public Roads 70 (5) (2007).
11. C. Jones, Perspective on freight congestion, Public Road, ProQuest Science Journals 36 (2007).
12. F. Glover, D. Klingman, N.V. Phillips, Network Models in Optimization and Their Applications in Practice, John Wiley & Sons, Inc., 1992
13. R.K. Ahuja, T.L. Magnanti, J.B. Orlin, Network Flows, Prentice Hall, New Jersey, 1993.

CITATION

Sutanto Soehodho, Nahry, Traffic flow consideration in design of freight distribution system, IATSS Research, Volume 34, Issue 1, July 2010, Pages 55-61, ISSN 0386-1112, http://dx.doi.org/10.1016/j.iatssr.2010.06.007.

CHAPTER 3

Real-Time Road Traffic Anomaly Detection

Jamal Raiyn[1], Tomer Toledo[2]

[1]Faculty of Exact Science, Computer Science Department, Al Qasemi Acedemic College, Baqa Al Qarbiah, Israel
[2]Faculty of Civil and Environmental Engineering, Department of Transportation Engineering, Technion, Haifa, Israel

ABSTRACT

Many modeling approaches have been proposed to help forecast and detect incidents. Accident has received the most attention from researchers due to its impacts economically. The traffic congestion costs billions of dollars to economy. The main reasons of major percentage of traffic congestion are the incidents. Road accidents continue to increase in digital age. There are many reasons for road accidents. This paper will discuss and introduce new algorithm for road accident detection. Various forecast schemes have been proposed to manage the traffic data. In this paper we will introduce road accident detection scheme based on improved exponential moving average. The proposed traffic incident detection algorithm is based on the automatic exponential moving average scheme. The detection algorithm is based on analyzing the collected traffic flow parameters. The detection algorithm is based on analyzing the collected traffic flow parameters. In addition a real-time accident forecast model was developed based on short-term variation of traffic flow characteristics.

INTRODUCTION

The main reason in accidents on the highway can be divided into four categories such as the environment, traffic conditions, vehicles and drivers behavior. Many studies [1] - [4] showed that higher speeds did not lead to serious accidents. On the other hand, some studies showed that fatal accidents increased with high speed limits. Our analysis reveals that the major factor leading to an accident is not speed itself but the variation of speed. There are three basic strategies to relieve congestion [5] : The first strategy is to increase the transportation infrastructure. However this strategy is very expensive and can only be accomplished in the long term. The second strategy is to limit the traffic demand or make traveling more expensive, which will be strongly disapproved of by travelers. The third strategy is to focus on efficient and intelligent utilization of the existing transportation infrastructures. This strategy is a best trade-off and gains more and more attention. Currently, the Intelligent Transportation System (ITS) is the most promising approach to implementation of the third strategy. Various forecast schemes [6] - [9] have been proposed to manage the travel flow information. Meanwhile the robustness and accuracy of the exponential smoothing forecast are high and impressive. This paper reports on the performance of three moving average techniques in predicting average travel speeds up to 10 minutes ahead of time. The advantage of the exponential smoothing algorithm is simple. However its forecast precision is not high. If a high forecast precision is requested, it is necessary to consider the real-time information includes the non-conditions events. This paper introduces road accident detection scheme. Road accident detection scheme is focused on real-time information. The real-time information has been achieved to update the historical adaptive information.

To optimize the detection algorithm we have collected travel data by the mobile phone. For a successful forecast of traffic flow, it ought to apperceive the variety of environment and can adjust the parameters automatically. Furthermore it is important that the forecast model takes into consideration the abnormal conditions that occur in real-time [4] [10] [11] .

The paper is organized as fellow: Section 2 describes the methodology of road accidents detection scheme. Section 3 and section 4 discuss the performance analysis of the proposed detection scheme and illustrate the simulation results.

METHODOLOGY

This section presents a methodology to detect road accidents based on travel time variations. We consider accident during peak periods (i.e., morning or afternoon) and during non-peak periods. The observed traffic data consists of normal and abnormal (accident) travel data. The abnormal record is at least 30 km/h lower traffic speed than the average speed of all records at the same time on the same day of the week. The threshold of 30km/h is a symbolic value of the smallest speed change that people would consider "abnormal". Threshold determination depends on the travel observation data. Equation (1) will be used to forecast the accident scheme.

$$tt(t+1,k,acci) = \alpha \times tt(t,k,acci) + (1-\alpha) \times EMA(t,k,acci) \tag{1}$$

Alpha can be expressed as follows:

$$\alpha = \frac{1}{1 + \left[\dfrac{Var(k)}{E(k)} \right]}$$

where Var(k) is the variance of the expected number of crashes at the reference sites. E(k) is the expected number of crashes at these reference sites.

Section Mutual Influence

In the real-time forecasting we take into consideration the effect of the upstream (UP) and downstream (DS) as illustrates in Equation (2).

$$tt(t+1,k) = tt^{H}(t+1,k) + \gamma_1 \times \text{desired} + \gamma_2 \times \text{UP} + \gamma_3 \times \text{DS} \tag{2}$$

where

$\text{desired} = \left[tt^{M}(t,k) - tt^{H}(t,k) \right]$

$\text{upstearm} = \left[tt^{M}(t,k-1) - tt^{H}(t,k-1) \right]$

$\text{downstream} = \left[tt^{M}(t,k+1) - tt^{H}(t,k+1) \right]$

$tt_{\text{abnormal}}(t,k) = tt^{-}(t-1,k) + tt^{+}(t+1,k)$

$$\delta = \Delta \big(tt(t,k) - tt(t+1,k) \big)$$

k is the desired section, (k − 1) is the upstream section, (k + 1) is the downstream section.

Figure 1 and Figure 2 illustrate the abnormal condition in the up and down stream.

Accident Detection Strategy

The performance of an incident detection system is determined on two levels: data collection and data processing. Data collection refers to the detection/sense/surveillance technologies that are used to obtain traffic flow data. Data processing refers to the algorithms used for detecting and classifying incidents through analyzing the traffic parameters from detectors or sensors for the purpose of alerting observers of the occurrence, severity, and location of an incident. The hybrid of data collection strategies and data processing methodologies results in a variety of solutions for incident detection. The main task of the proposed accident detection (AD) algorithm is to identify and distinguish different traffic modes in Table 1. It depends on an upstream occupation increase and a downstream occupation decrease at the level of loop detector where an incident happened. This algorithm compares a value of a traffic flow parameter with a known value. The algorithm trusts that an upstream occupation will increase and downstream occupation will decrease where an incident happened. In traffic incident detection, a time sequence is used to describe a traffic state. When a current measured value is deviated from the output of the algorithm seriously, the algorithm will think that an incident has occurred. The time sequence analytic algorithms include a moving average algorithm, an exponential smoothing algorithm.

- The accident characterized by temporal variation of speed at fixed road section (location) expressed as the coefficient of variation in speed.
- The spatial variation of speed along road sections expressed as the difference in speed between upstream and downstream location (Q).

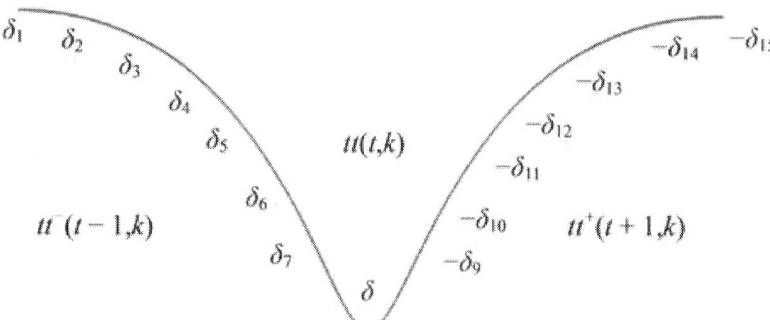

Figure 1. Variation of the travel time speed.

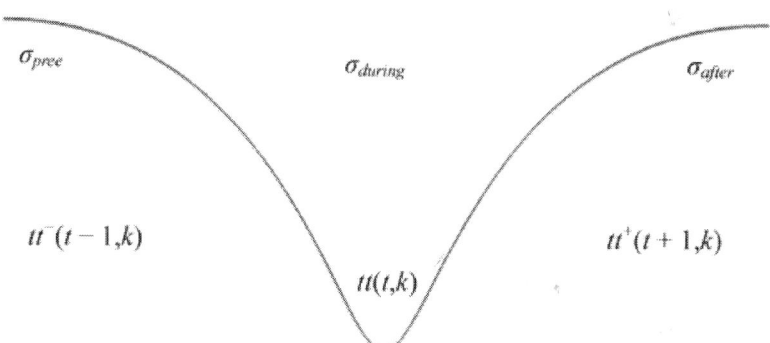

Figure 2. Accident characteristics-stdev.

Table 1. Optimized parameters in AD/NAD.

	AD	no AD
γ	0.9993	0.5346
β_1	0.9081	0.2215
β_2	0.9834	0.1138
β_3	0.8591	0.2315
β_4	0.9993	0.4643

$$Q = \left| \tilde{tt}\,(t,s1) - \tilde{tt}\,(t,s2) \right|$$

(3)

Where $\tilde{u}(t,s1)$, $\tilde{u}(t,s2)$ average speeds computed over period of t upstream and downstream of a road sections, respectively (km/h).

Incident-Influence Traffic Data

An incident occurring on section i within time interval t is considered to have a significant impact on traffic when traffic measurements from the upstream and downstream stations satisfy the following conditions:

1) The difference between upstream speed si, t and downstream speed si + 1, t is greater than the threshold value;
2) The ratio of the difference between the upstream and downstream speeds to the upstream speed (si, t ? si + 1, t/si, t, is greater than the threshold value;
3) The ratio of the difference between the upstream and downstream speeds to the downstream speed (si, t − si + 1, t)/si + 1, t is greater than the threshold value.

The abnormal record shows that at least 30 km/h lower traffic speed than the average speed of all records at the same time on the same day of the week. The threshold of 30 km/h is a symbolic value of the smallest speed change that people would consider "abnormal". The vehicle speed starts to decrease in upstream however the speed in downstream starts to increase. When an incident occurs between stations k and k + 1, the congestion causes a clear difference between the occupancies of the upstream and the downstream stations as illustrates Figure 3.

$$\frac{tt(k,t)-tt(k+1,t)}{tt(k,t)} > \text{threshold}$$

(4)

$$\frac{tt(k,t))-tt(k+1,t)}{tt(k+1,t)} > \text{threshold}$$

(5)

$$\text{Mean(accidents)} = \frac{1}{N}\sum_{i=1}^{n}(\mu-\sigma_i)$$

(6)

σ standard deviation, N number of the acidents.

Real-Time Accident Detection

The travel time forecast model considers the incident and non-incident conditions. We make different between:

- Accident during peak time (morning/afternoon);
- Accident during regular time;
- Heavy accident;
- Light accident.

The accident is cleared at current time t in section s, the duration is known and the speed is considered to be 30 km reduced of the average speed.

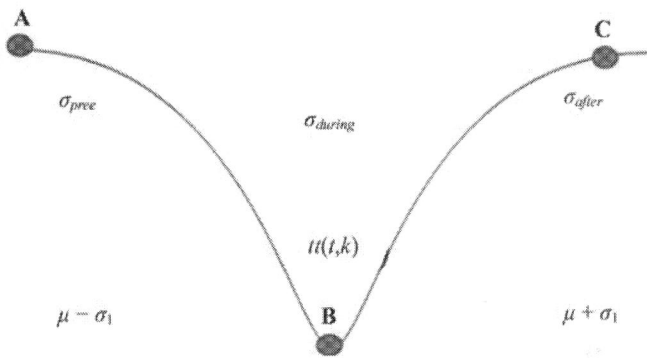

Figure 3. Accident characteristics of StDev.

$$tt\left(t+1,k\right)=tt^{H}\left(t+1,k\right)+\gamma\times\left(P_{t}\right)\times\left(tt_{t}^{M}-tt_{t}^{H}\right)$$

$$P_{t}=P\left(\text{accident}\right)_{t}=\frac{1}{1+e^{-v_{t}}},$$

$$x_{1}=\frac{\left(\sigma_{t}-\sigma_{t}^{H}\right)}{\sigma_{t}^{H}},$$

$$x_{3}=\frac{\left(tt_{t}-tt_{t}^{H}\right)}{\sigma_{t}^{H}}-\frac{\left(tt_{t-1}-tt_{t-1}^{H}\right)}{\sigma_{t-1}^{H}}$$

where X denotes the vector of predictor variables. β is the vector of coefficient associated with the predictor variables. and can be computed according to the binary logit model. v_t is the logit link function (which is a linear combination of the predictor variables).

Accident Probability

Based on statistical measurements of historical information and real information, the forecast model can estimate the occurrence of abnormal conditions without external information as express Equation (7) and Equation (8).

$$tt_{acc}^{F}\left(t+1,k\right)=EMA_{acc}^{H}\left(t+1,k\right)+\delta\left(tt_{acc}^{M}\left(t,k\right)-tt_{acc}^{H}\left(t,k\right)\right)$$

$$(7)$$

$$tt_{acc}^{F}\left(t+1,k\right)=EMA_{acc}^{H}\left(t+1,k\right)+\underbrace{\delta\left(tt_{acc}^{M}\left(t,k\right)-tt^{H}\left(t,k\right)\right)}_{\text{Corrections needed=epsilon}}$$

$$(8)$$

where,

$$\varepsilon=\left(\text{mean}\left(EMA_{\text{normal}}\left(k,t\right)\right)-\text{mean}\left(EMA_{\text{abnormal}}\left(k,t\right)\right)\right)$$

$$P_{\text{accident}}^{\text{hist}}\left(tt\left(t,k\right)\right)=P\left\{\sigma\left(tt_{i}\left(t,k\right)\right)>\delta_{th}\left(\text{hist}\right)\right\}$$

$$(9)$$

$$P_{\text{accident}}^{\text{real}}\left(tt\left(t+1,k\right)\right)=P\left\{\sigma\left(tt_{i}^{M}\left(t,k\right)\right)>\text{max}_{i}\,\sigma\left(tt_{i}^{\text{hist}}\left(t,k\right)\right)\right\}$$

$$(10)$$

where:

$\delta_{th2}\left(M\right)>\delta_{th1}\left(\text{hist}\right)$

$\sigma\left(tt^{M}\left(t,k\right)\right)>\sigma\left(tt^{M}\left(t-1,k\right)\right)$

$tt^{M}\left(t,k\right)>tt^{H}\left(t,k\right)$

$tt^{M}\left(t,k\right)>tt^{H}\left(t-1,k\right)$

The Total number of the expected accident is expressed as following:

$$E\left(N^{M}_{\text{accident}}\right)=\left(1-P^{\text{real}}_{\text{accident}}\right)\times n$$

(9)

Smoothed Parameter Optimization

To increase the exponential moving average forecast accuracy in real-time, the smoothed parameter alpha and gamma in Equation (4) should be optimized. Figure 4 illustrated the value of the optimized smoothed parameter gamma in real-time accident conditions.

Figure 4 and Figure 5 illustrate values of the optimized smoothed parameter gamma in real-time accident and non-accident conditions in highway. However Figure 6 and Figure 7 illustrate values of the optimized smoothed parameter gamma in real-time accident and non-accident conditions in urban road.

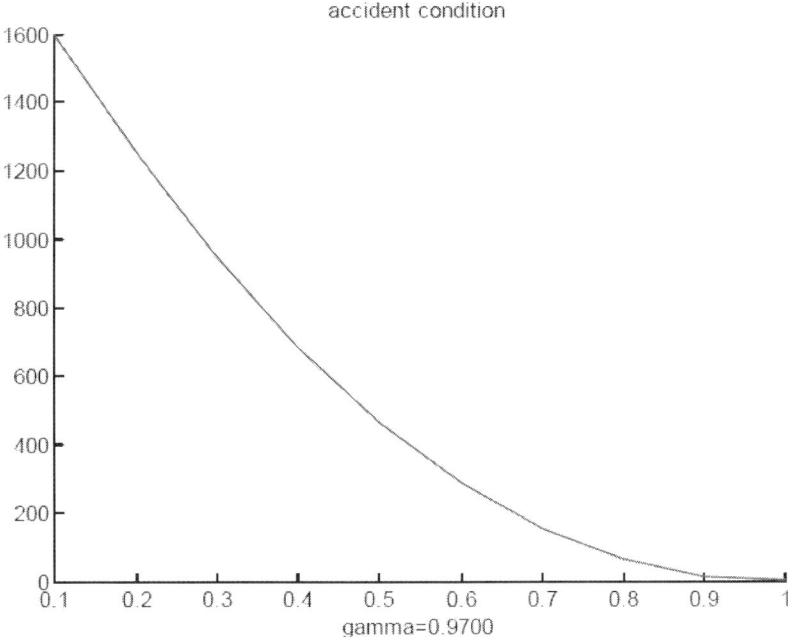

Figure 4. AC in highway.

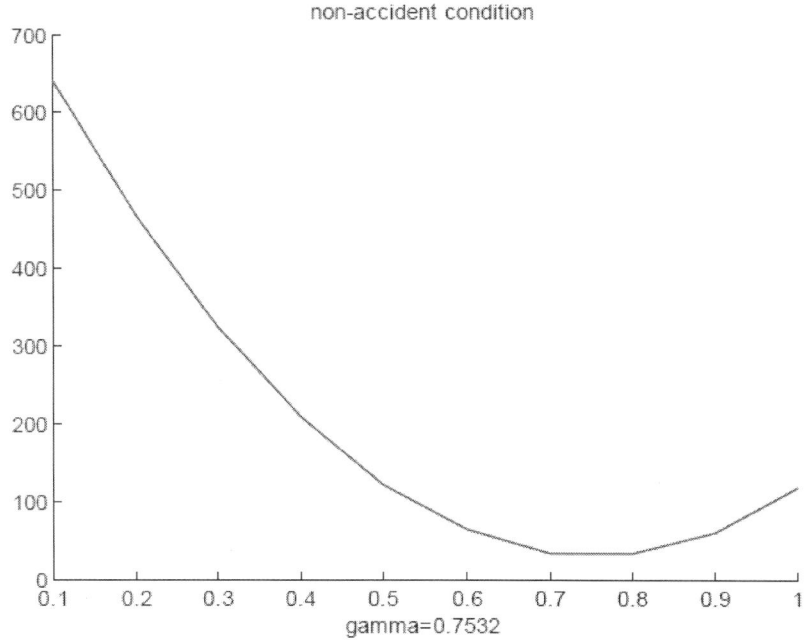

Figure 5. NAC in highway.

PERFORMANCE ANALYSIS

There are various measures of forecasting accuracy techniques proposed in the literature [5] [12] -[15] . The aim of this study is to evaluate forecast accuracy travel observations. The forecasting accuracy techniques are used to be able to select the most accurate forecast scheme. The forecasting performance of the various models and the measures of the predictive effectiveness was evaluated using various summary statistics. The comparing experiments are carried out under normal traffic condition and abnormal traffic condition to evaluate the performance of four main branches of forecasting models on direct travel time data obtained by license plate matching (LPM). The MAE is a measure of overall accuracy that gives an indication of the degree of spread, where all errors are assigned equal weights. The MSE is also a measure of overall accuracy that gives an indication of the degree of spread, but here large errors are given additional weight. It is the most common measure of forecasting accuracy.

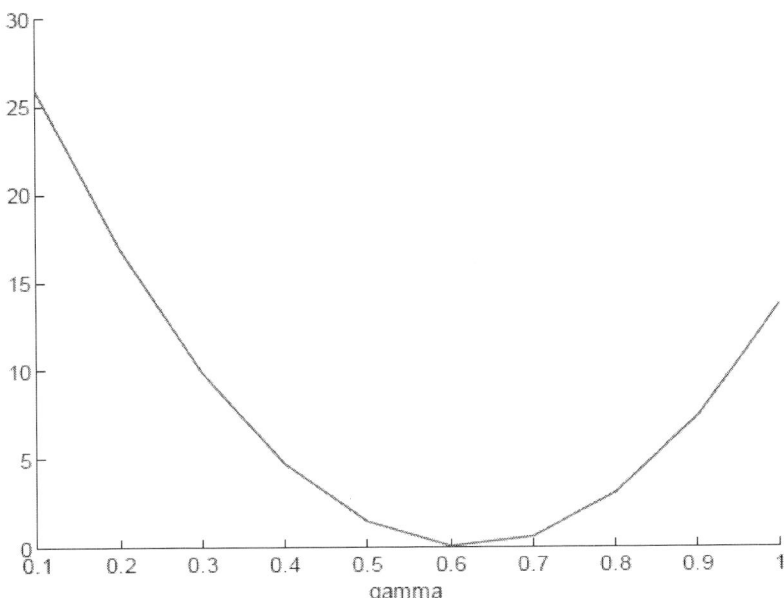

Figure 6. NAC in urban road.

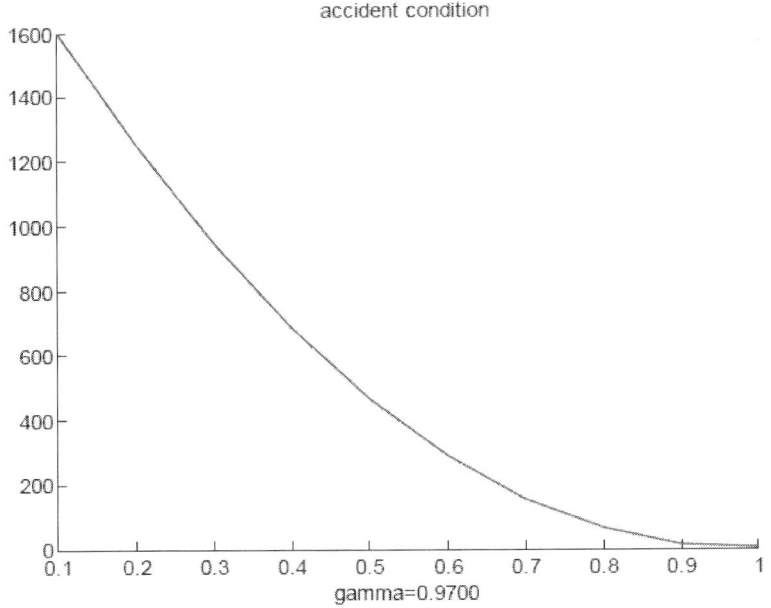

Figure 7. AC in urban road.

Often the square root of the MSE, RMSE, is considered, since the seriousness of the forecast error is then denoted in the same dimensions as the actual and forecast values themselves. Mean square percentage error (MSPE) is the relative measure that corresponds to the MSE. The more commonly used measure is the root mean square percentage error (RMSPE). Theil's Coefficient is another statistical measure of forecast accuracy. One specification of Theil's compares the accuracy of a forecast model to that of a naive model. A Theil's greater than 1.0 indicates that the forecast model is worse than the naïve model; a value less than 1.0 indicates that it is better. The closer U is to 0, the better the model.

SIMULATION RESULTS

The travel observation data consists of normal and abnormal (accident) travel data. Figure 8(a) and Figure 8(b) illustrate the abnormal conditions in up and download stream in peak hours. However Figure 8(c) illustrates the abnormal condition in no peak hours.

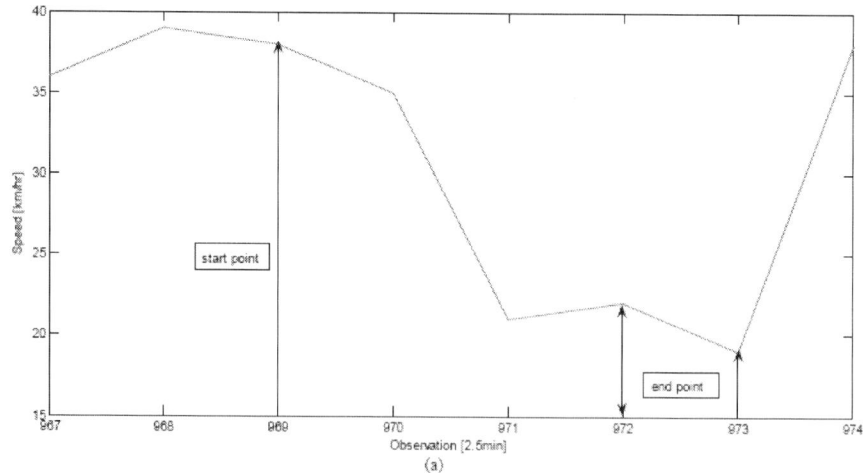

(a)

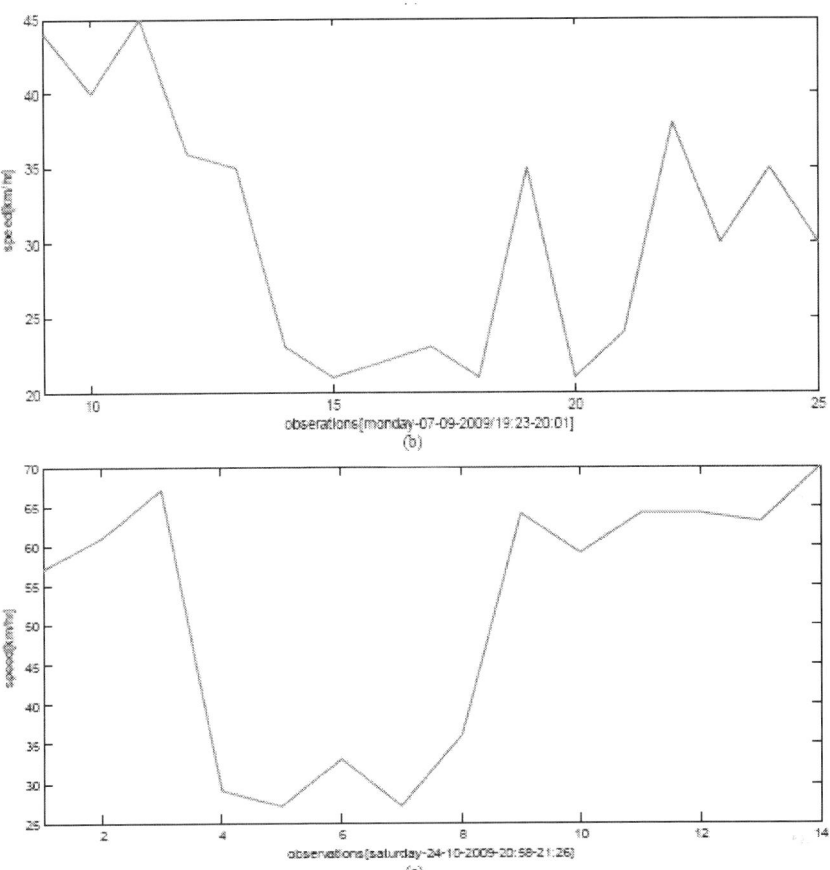

Figure 8. (a) Travel time variation in AC; (b) Travel time variation in AC; (c) Travel time variation in AC.

Table 2 and Table 3 illustrate the performance analysis of exponential moving average scheme based on historical and real time forecasting. The comparison has been introduced based on accident and non accident conditions.

Table 4 describes the comparison of exponential moving average scheme based on sorted data that the difference between two neighbor observations is bigger than 5 km and 10 km. Figure 9 illustrates the comparison between exponential moving average and improved exponential moving average.

Table 2. Hist vs. real-time in NAC.

Non-Accident Condition	Hist	Real
mean data	67.805	67.805
mean prediction	65.622	66.798
std data	17.809	17.809
std prediction	18.682	16.968
Observations with error over 5 km/hr	33.086	31.293
Observations with error over 10 km/hr	17.385	15.735
max abs. error	73.39	73.264
max relative error	587.12	586.11
mean error	2.183	1.0076
mean abs. error	6.6768	5.472
mean relative error	12.238	10.562
root mean squared error	12.452	9.2418
root mean squared percent error (1)	26.42	23.514
root mean squared percent error (2)	18.364	13.63
Theil's coefficient	9.0011	6.6476
bias proportion	3.0737	1.1886
variance proportion	0.49122	0.82716
co-variance proportion	96.435	97.984

Table 3. Hist vs. real-time in AC.

Accident Condition	Hist	Real
mean data	79.234	79.234
mean prediction	75.324	78.981
std data	17.737	17.737
std prediction	22.993	16.673
Observations with error over 5 km/hr	42.206	40.281
Observations with error over 10 km/hr	26.006	22.356
max abs. error	93.492	81.288
max relative error	1181.7	4538.8
mean error	3.9104	0.25324
mean abs. error	10.588	7.4191
mean relative error	16.743	12.656
root mean squared error	20.505	12.118
root mean squared percent error (1)	39.798	32.049
root mean squared percent error (2)	25.88	15.294
Theil's coefficient	12.82	7.4844
bias proportion	3.6367	0.04367
variance proportion	6.57	0.77046
co-variance proportion	89.793	99.186

Table 4. Up- and downstream effect.

Real-time	EMA	Speed > 5 km	Speed > 10 km
mean data	69.276	58.275	56.729
mean prediction	55.884	49.321	49.265
std data	23.449	21.738	20.014
std prediction	22.77	4.1422	2.2321
Observations with error over 5 km/hr	92.415	85.318	83.302
Observations with error over 10 km/hr	78.546	71.246	67.392
max abs. error	86.853	69.984	64.842
max relative error	408.87	520.96	527.45
mean error	13.392	8.9542	7.4641
mean abs. error	15.299	19.812	17.62
mean relative error	25.257	38.407	35.075
root mean squared error	18.179	23.746	21.341
root mean squared percent error (1)	32.904	54.475	50.779
root mean squared percent error (2)	26.242	40.747	37.619
Theil's coefficient	13.619	21.26	19.495
bias proportion	54.27	14.219	12.233
variance proportion	0.13941	54.909	69.425
co-variance proportion	45.591	30.872	18.342

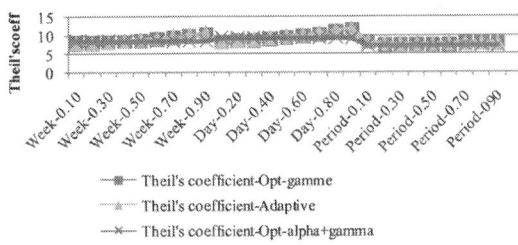

—■— Theil's coefficient-Opt-gamme
········ Theil's coefficient-Adaptive
—×— Theil's coefficient-Opt-alpha+gamma

Figure 9. EMA vs. improved EMA.

CONCLUSION

Analysis of the road incidents based on the speed variation is not robust enough to develop real-time forecast model. Because a speed observation can be zero when there is no vehicle, or the system collects a wrong speed observation, in this case, the computation of CVS can be done in many variations.

REFERENCES

1. Ronen, B., Coman, A. and Schragenheim, E. (2001) Peak Management. International Journal of Production Research, 39, 3183-3193.http://dx.doi.org/10.1080/00207540110054588

2. Tu, H., Van Lint, H. and Van Zuylen, H. (2008) The Effects of Traffic Accidents on Travel Time Reliability. IEEE Conference on Intelligent Transportation Systems, Beijing, 12-15 October 2008.

3. Wang, Z. and Murray-Tuite, P. (2010) Modeling Incident-Related Traffic and Estimating Travel Time with a Cellular Automaton Model. Proceedings of Transportation Research Board's 89th Annual Meeting CD-ROM, 10-14 January 2010, Washington, DC.

4. Wild, D. (1997) Short-Term Forecasting Based on a Transformation and Classification of Traffic Volume Time Series. International Journal of Forecasting, 13, 63-72.http://dx.doi.org/10.1016/S0169-2070(96)00701-7

5. Zheng, X. and Liu, M. (2009) An Overview of Accident Forecasting Methodologies. Journal of Loss Prevention in the Process Industries, 22, 484-491.http://dx.doi.org/10.1016/j.jlp.2009.03.005

6. Andrada-Felix, J. and Fernandez-Rodriguez, F. (2008) Improving Moving Average Trading Rules with Boosting and Statistical Learning Methods. Journal of Forecasting, 27, 433-449. http://dx.doi.org/10.1002/for.1068

7. Guin, A. (2006) Travel Time Prediction Using a Seasonal Autoregressive Integrated Moving Average Time Series Mode. Proceedings of the IEEE Intelligent Transportation Systems Conference, Toronto, 17-20 September 2006, 493- 498.

8. Lv, Y. and Tang, S. (2010) Real-time Highway Traffic Accident Prediction Based on the K-Nearest Neighbor Method. International Conference on Measuring Technology and Mechatronics Automation, Volume 3, 547-550.

9. Xia, J (2010) Predicting Freeway Travel Time under Incident Condition. Proceedings of Transportation Research Board's 89th Annual Meeting CD-ROM, 10-14 January 2010, Washington, DC.

10. Alger, M. (2004) Real-Time Traffic Monitoring Using Mobile Phone Data. Proceedings of 49th European Study European Study Group with Industry, Oxford, United Kingdom.

11. Stephanedes, Y.J., Michalopoulos, P.G. and Plum, R.A. (1981) Improved Estimation of Traffic Flow for Real-Time Control. Transportation Research Record, 795, 28-39.

12. Jo, H., Lee, B., Na, Y.-C., Lee, H. and Oh, B. (2007) Prioritized Traffic Information Delivery Based on Historical Data Analysis. Proceedings of the 2007 IEEE Intelligence Transportation Systems Conference, Seattle, September 30-Oc- tober 3 2007, 568-573.

13. Karim, A. and Adeli, H. (2003) Fast Automatic Incident Detection on Urban and Rural Freeways Using Wavelet Energy Algorithm. Journal of Transportation Engineering, 129, 57-68. http://dx.doi.org/10.1061/(ASCE)0733-947X(2003)129:1(57)

14. Lee, H., Chowdhury, K.N. and Chang J. (2008) A New Travel Time Prediction Method for Intelligent Transportation Systems. Springer-Verlag, Berlin, 473-483.

15. Xiaoqiang, Z., Ruimin, L. and Xinxin, Y. (2010) Incident Duration Model on Urban Freeways Based on Classification and Regression Tree. 2nd International Conference on Intelligent Computation Technology and Automation, TRB 2010 Annual Meeting, 2, 526-528.

CITATION

Raiyn, J. and Toledo, T. (2014) Real-Time Road Traffic Anomaly Detection. *Journal of Transportation Technologies*, **4**, 256-266. doi: 10.4236/jtts.2014.43023.

CHAPTER 4

Motion Planning of Autonomous Vehicles on a Dual Carriageway without Speed Lanes

Rahul Kala and Kevin Warwick

School of Systems Engineering, University of Reading, Whiteknights, Reading, Berkshire RG66AY, UK;

ABSTRACT

The problem of motion planning of an autonomous vehicle amidst other vehicles on a straight road is considered. Traffic in a number of countries is unorganized, where the vehicles do not move within predefined speed lanes. In this paper, we formulate a mechanism wherein an autonomous vehicle may travel on the "wrong" side in order to overtake a vehicle. Challenges include assessing a possible overtaking opportunity, cooperating with other vehicles, partial driving on the "wrong" side of the road and safely going to and returning from the "wrong" side. The experimental results presented show vehicles cooperating to accomplish overtaking manoeuvres.

INTRODUCTION

Autonomous vehicles [1,2] are capable of driving themselves in traffic scenarios and are seen as a replacement for human-driven vehicles in the future. They make transportation systems efficient and safe [3,4] and are therefore sources of research and development. The problem of planning deals with decision making regarding the motion of a vehicle. Broadly, planning is responsible for the trajectory generation and speed computation of the vehicle at any instance of time. The planning framework enables vehicles to judiciously avoid all obstacles and other vehicles, in cooperation with each other. Current planning algorithms [5,6,7] largely assume predefined speed lanes within which the vehicles need to drive. Planning is hence mostly restricted to deciding the judicious lane and speed of travel, while making travel efficient and safe.

The notion of planning in the absence of speed lanes is motivated by traffic systems in countries where speed lanes are not followed and vehicles can drive in-between what would normally be the predefined speed lanes. Such traffic is unorganized and, at times, chaotic. The details of such systems can be found in [8,9]. Consider the case when traffic has a high diversity in terms of vehicle sizes. Therefore, narrower vehicles, occupying a complete lane, effectively leave unused road width, which can therefore be utilised to accommodate additional vehicles, if the traffic is unorganized. This leads to unorganized traffic, resulting in a higher traffic bandwidth when compared to its organized counterpart over the same width of road. Further, when considering traffic exhibiting a high diversity in the speed of vehicles, it would be problematic for a high speed vehicle to have to follow a very slow vehicle on the road, and therefore, it would be best for an overtaking opportunity to be arranged as quickly as possible. By planning in terms of unorganized traffic, vehicles can be spread across the road, thereby overcoming lane issues and resulting in making overtaking quite feasible, whereas it would not have been otherwise.

Hence, unorganized traffic can be more efficient in scenarios where vehicles vary largely in their speed capabilities and size. However, such traffic is more risky and is more likely to lead to accidents due to the unclear intentions of the vehicles around [10,11]. Indian traffic is a clear example where vehicle sizes vary from two-wheeled motorbikes and three-wheeled auto rickshaws, to buses and trucks. Speeds vary considerably between manually-driven vehicles and cars. Traffic is unorganized, and vehicles cut in whenever they find space. Constant overtaking manoeuvres are visible. There is a likelihood of organized traffic taking the shapes of unorganized traffic with the introduction of autonomous vehicles, which vary in speeds and sizes.

A common characteristic of traffic systems in many countries with narrow roads is that there is no physical barricade for inbound and outbound traffic on a dual carriageway. The road may be divided by markers or drivers may assume that half of the road is for inbound traffic and the other half for outbound traffic. Hence, they stick to their own side, normally following the vehicle ahead. In an unorganized traffic scenario, there is, though, the chance of overtaking for motorcycles and smaller vehicles. There has been very little research in the domain of intelligent vehicles for dual carriageways, while they constitute an important aspect of traffic.

In an earlier work by the authors [12], the task of motion planning for autonomous vehicles in the absence of speed lanes was investigated. The assumption was, however, that the entire traffic flow was in one direction (outbound or inbound). With this assumption, the algorithm could present interesting vehicular behaviours in complex scenarios. The assumption is not difficult, since a road can always be assumed to have a virtual boundary dividing the two directions of travel. Given this virtual boundary, for planning, each side can display its own behaviours, and the two sides do not need to interact with each other at any point of time.

The use of such behaviour-based systems is common in mobile robotics [13], where a variety of behaviours are designed based on different scenarios and a framework integrates the different behaviours for the optimal motion of the robot. Dee and Hogg [14] studied the robot behaviours in light of the general navigation of humans, and highlighted navigation using the shortest path or the simplest path with the least number of turns. Even autonomous vehicles model and integrate different possible behaviours. For organized travel, a limited range of behaviours are possible, as demonstrated in [15,16,17].

This work is focussed on extending and generalizing the solution [12] for performance, wherein both inbound and outbound traffic operate on the same road without any physical barricade in between. The focus here is hence to only study behaviours where vehicles travelling in the opposite direction interact in some way or the other. General travel, when vehicles remain on their own side, is exactly as one would expect with the earlier approach and, hence, is not covered in this work.

This raises the important question: should vehicles be allowed to slip across to the "wrong" side of a dual carriageway with un-barricaded inbound and outbound traffic, for some time? In general, this is not regarded as safe, even for human drivers, because a driver slipping over to the wrong side may not be able to return back to the correct side and might therefore cause an accident or traffic jam. Hence, for non-autonomous traffic, even if it appears that for a vehicle to occupy some part of the road on the opposite side, this would lead to better traffic bandwidth and travel efficiency, it must be avoided at all costs due to the risks involved. Unfortunately, this eliminates much of the possible interesting behaviour involving the mixing of traffic from opposite directions.

However, it is common for a vehicle to slip into the wrong side just to overtake. Overtaking is the key factor contributing to efficiency in diverse speed unorganized traffic. Hence, every attempt is made to enable a faster vehicle to overtake a slower vehicle. If a faster vehicle considers it safe enough, it should therefore be allowed to slip into the wrong side, complete the overtaking manoeuver and return to the correct side whenever feasible. Such overtaking may thereby greatly enhance travel efficiency, while making it a little riskier for the traffic travelling on the opposite side of the road if the assessment of the overtaking vehicle is poor. In this paper, such overtaking is modelled as a single-lane overtake behaviour. This behaviour joins the pool of behaviours modelled in [12].

The single-lane overtake is largely inspired by narrow road traffic systems with one lane per side of travel. In such traffic systems, the addition of a slow vehicle can almost block a complete road. Hence, it is important for an autonomous vehicle to have some way of overtaking and allowing itself and the other vehicles to drive efficiently. Even in countries with organized traffic, human drivers tend to take any opportunity to overtake in such scenarios. On many occasions, all vehicles collectively decide a strategy. A lane-following law-abiding autonomous vehicle can be troublesome in such situations, if such a behaviour is not modelled. Another source of motivation is obtained from zones in traffic systems, which a vehicle may use for overtaking. Similarly, such behaviours are common in countries with unorganized traffic and are usually taken with great caution.

Overtaking, even in general, is regarded as a special behaviour and has been extensively studied in the literature, including overtaking assessment and actually performing overtaking. Overtaking involves a change of lane to the overtaking lane, driving ahead of the vehicle to be overtaken and then returning to the driving lane. Overtaking in organized traffic is easier to carry out as a set of lane changes, because the driving lane and the overtaking lane are on the same side and in the same direction of travel, so this minimizes the risks involved.

The contributions of this research, which complement the contributions of prior work [12], are:

- The problem of motion planning for autonomous vehicles in a dual carriageway setting and working without communication is studied. This is a problem that has been relatively un-touched in the literature.
- Single-lane overtaking behaviour is modelled and studied, while the literature largely studies overtaking when both the normal lane and overtaking lane have traffic flowing in the same direction.
- Single-lane overtaking behaviour is modelled in a prioritized set of behaviours.
- A mechanism for cancelling single-lane overtaking is designed, while the literature normally assumes that every overtaking that is initiated must complete successfully.

This paper is organized as follows. Section 2 summarizes some related research works. Section 3 summarizes the directly relevant previous work [12], which is extended in this paper. Section 4 presents the single-lane overtaking behaviour. Experimental results are then shown in Section 5, and conclusions are given in Section 6.

RELATED WORKS

We omit here a detailed discussion of the related issues in light of the literature studies, for which readers are referred to [12]. A few interesting recent works are however discussed here. We first discuss the works related to overtaking, although there are marked differences with the single-lane overtaking behaviour studied here and the typical overtaking problem by legal lane changes as studied in the literature. Naranjo *et al.* [18] framed different rules for a vehicle departing from its lane and joining the overtaking lane, motion in the overtaking lane and returning to the original lane of travel. Similarly, Jin-ying et al. [19] performed a fuzzy modelling of the overtaking procedure and defined separate fuzzy membership functions for each of the stages, based on which, a fuzzy inference was done for motion control. Petrov and Nashashibi [20] designed an adaptive controller to perform the different stages of overtaking.

Hegeman *et al.* [21] developed an assistance system for drivers. The authors assessed the feasibility of overtaking and only overtaking that exceeded a safety threshold was allowed. Similarly, Wang *et al.* [22] used uncertainty modelling to compute the probability of a collision between vehicles, based on which, the decision of whether to overtake or not was made. Karaduman *et al.* [23] used optical flow to derive the distances from the vehicles and used Bayesian belief networks to compute the overtaking risk in a three-vehicle scenario. These approaches did not simultaneously check for the feasibility of and actually carry out overtaking and did not allow for an emergency cancellation of overtaking. Further, any cooperation between the vehicles was not modelled.

A major problem in overtaking is the lack of information due to occlusion by the vehicles directly in front, which restricts visibility and, hence, the ability to make judicious decisions. Olaverri-Monreal *et al.*[24] proposed the use of vehicle communication to provide enhanced visibility to the user, wherein a vehicle may communicate visual information to another vehicle for decision making. Similarly, Milanés*et al.* [25] built a vision system to identify the positions and speeds of the vehicles using stereovision, the information about which was used by a controller to carry out the overtaking procedure.

Kuwata *et al.* [26] used rapidly-exploring random trees (RRT) for the generation of the trajectory of a single vehicle. The sampling was made biased for early generation of good results. In a related work, Gehrig and Stein [27] used elastic bands to model the vehicle following behaviour. The band was attached to the vehicle being followed, and the elastic band could model dynamic obstacles in the following procedure. Kala and Warwick [28] proposed a hierarchical distribution of the problem of motion planning for multiple autonomous vehicles with communication. The layers consisted of route planning, a static obstacle avoidance layer, a vehicle coordination layer and a low level trajectory generator. A cooperative overtaking behaviour for lane-based traffic is shown in the work of Frese and Beyerer [29]. Here, the authors studied mixed integer programming, tree search, elastic bands, random priorities and optimized priorities as underlying algorithms for trajectory generation. Meanwhile, Sewall*et al.* [30] used a prioritized A* algorithm for the motion of vehicles. The map was divided into segments.

Chu *et al.* [31] used a fixed set of waypoints to construct a limited collection of candidate paths with a limited set of manoeuvers. The best path was selected out of these and used for vehicle navigation. Limited waypoints do enable fast trajectory computation; however, the optimal trajectory may be missed. Anderson *et al.* [32] used constrained Delaunay triangles to compute a vehicle's trajectory. The resultant trajectory was short in length and benefitted from the vehicle being clearly separated at all times from obstacles. In both approaches, the method of treating other vehicles as static obstacles did though produce a lack of cooperation between all vehicles. Further, sudden steering (lane changes) of a slow vehicle in front could cause a collision with a fast vehicle to the rear (which was initially travelling in a separate lane).

Kala and Warwick [33], under the assumption of one way traffic only, performed motion planning amongst multiple vehicles. Different methods of obstacle and vehicle avoidance were tried, and the best path was optimized using the elastic strip algorithm. The algorithm was cooperative, however computationally expensive in the case of a large number of vehicles or obstacles. Sezer and Gokasan [34] carried out the planning of a ground vehicle by making it constantly move in the largest visible gap. Current orientation and non-holonomic constraints were considered in the choice of gap. The algorithm was, however, not cooperative in the presence of multiple vehicles. Unfortunately, a (normal) bounded road structure, unlike the experimented unbounded maps, can result in the vehicle being trapped between an obstacle and a road boundary in a narrow passage.

The problem of vehicle planning in unorganized traffic takes some characteristics from the problem of robot motion planning, and hence, some works from the domain are discussed. Kala [35] used cooperative coevolution for planning robots in narrow corridors. The corridors could occupy a single robot. The planning was centralized and assumed communication between the robots. Ryan [36] identified different structures in the robot map, like stacks, halls, cliques and rings, which served as subgraphs for motion planning. Each subgraph had a limited set of operations or behaviours. The proposed approach relies on one such structure, while a rich set of behaviours is presented.

Rashid *et al.* [37] used the concept of velocity obstacles for robot motion. They considered orientations of the robots to compute the velocity to avoid collisions. Daniel *et al.* [38] solved the problem of the sub-optimality of the A* algorithm due to the discretization of the graph by computing non-neighbouring grid expansions during a node expansion. Sgorbissa and Zaccaria [39] carried a coarser level planning using Voronoi graphs, and the robot was moved at the finer level using the potential-based approach. The robot was prevented from being struck in-between obstacles by identifying scenarios called roaming trails. The approach was non-cooperative. A solution to the general problem of multi-agent planning with bounded communication is presented in Wu *et al.* [40] using a decentralized, partially-observable Markov decision process.

MOTION PLANNING WITHOUT SPEED LANES

This section summarizes the work reported in [12]. The current work retains the same problem definitions, assumptions and solution framework. Given a traffic scenario with a number of autonomous and non-autonomous vehicles, the problem was to decide the motion of the autonomous vehicle being planned. A vehicle was assumed to be a rectangle of length (len_i) and width (wid_i). The current speed v_i was limited to a maximum value of $vMax_i$, while instantaneous accelerations were limited to the interval [$-accMax_i$, $accMax_i$]. The road segment was assumed to be bounded by sensed boundaries, called $Boundary_1$ and $Boundary_2$. The paper assumed a fully-known environment. Whilst this is an extensive assumption, it is possible with modern technology, with sensors, such as 3D LiDAR, ultrasonics, arrays of 3D vision cameras with advances in computer vision, *etc.* The possibility of having multiple intelligent vehicles in the vicinity also allows for the sharing of vision information across vehicles to rectify errors and give information about occluded areas. The algorithm used a road coordinate axis system along with the Cartesian coordinate axis system. The road coordinate axis system (X'Y') employed the X' axis as*Boundary₂* and the Y' axis as the ratio of the distance of the present point of the vehicle from *Boundary₂*, as compared to the overall road width. The representation of any point $P(x,y)$ could hence be given by Equation (1), and the terms are explained by Figure 1.

$$P(x,y) = P(x',y') = P\left(x', \frac{\|P(x,y) - Boundary_2(x')\|}{\|Boundary_1(x') - Boundary_2(x')\|}\right)$$

(1)

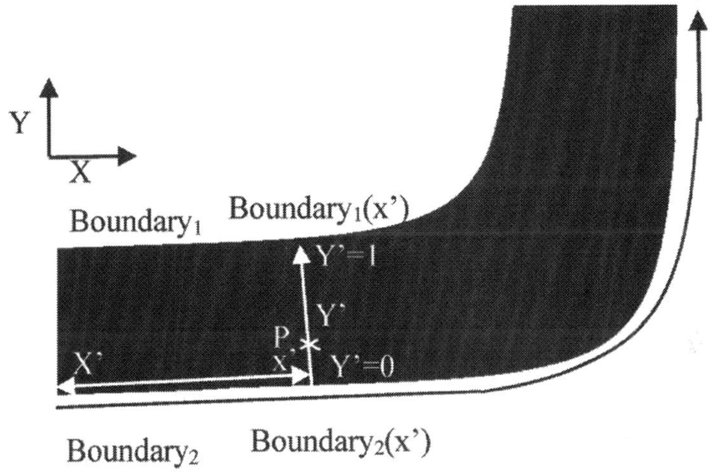

Figure 1. Road and Cartesian coordinate axis system [12].

The algorithm enabled planning and moving a vehicle, so that the separation available from the vehicle on all sides was larger than *separMinᵢ*. This was referred to as the aggression factor. The notion was to always keep as large a separation as possible, which meant greater safety; however, the attempt to maximize separation was subject to a threshold of *separMaxᵢ*. Higher values of the factor encouraged vehicles to make driving decisions, such that very high separation from other vehicles was very likely, while a low value of the factor encouraged aggressive driving decisions, such as close overtaking, wherein separations from other vehicles was likely to be very low. The solution was modelled as a set of behaviours. Each behaviour had a set of preconditions that had to be met for the particular behaviour to be initiated. The following is the complete list of behaviours:

- Obstacle avoidance: Obstacle avoidance behaviour is called whenever any obstacle or set of obstacles are sensed on the road. The vehicle attempts to overcome each obstacle by the widest possible margin (under a maximum of $separMax_i$).
- Centring: In the case that the vehicle does not find any obstacle or any other vehicle in the scenario, the vehicle slowly drifts towards the centre of the road.
- Lane change: Whenever any vehicle makes a lateral change on the road, the change carries the risk of an accident with any vehicle to the rear, which may be travelling at a higher speed and may now suddenly find that the vehicle in question has cut-in ahead of it. Hence, all behaviours involving changing of lateral positions are checked, and only those behaviours that ensure that all vehicles would have enough time to adjust their speeds are allowed, if the lane change behaviour actually proceeds.
- Overtaking: If a faster vehicle sees a slower vehicle directly ahead, which restricts its motion, it may attempt to overtake the slower vehicle. Overtaking can be direct or indirect. In direct overtaking, enough separation is available (along with the minimum safety distance) on either side of the slower vehicle, for overtaking to be initiated. If both sides can host the overtaking, the easier side is chosen. In indirect overtaking, enough separation is not initially available, but can be made available if the entire stream of vehicles ahead moves laterally on the road. Indentation.
- Be overtaken: If a slower vehicle ahead sees a faster vehicle behind trying to overtake, it may need to cooperate to give some additional separation to enable the overtaking to occur safely. The slower vehicle firstly assesses the side by which the overtaking is being attempted and then computes the distance by which it must laterally move.
- Maintain separation steer: This behaviour attempts to maximize the immediate separation available between the vehicle and any other

entity, if the immediate separation on either side is less than the maximum threshold of *separMax_i*.

- Slow down: If the vehicle cannot maintain the minimum separation of *separMin_i* on both sides, it is taken as a risk, and the vehicle is required to slow down.
- Discover conflicting interests: If two vehicles are seen to be steering towards each other because of any behaviours, there is clearly a potential risk. Once it is evident that a minimum threshold separation would be crossed, the trajectories of travel of the vehicles are straightened.
- Travel straight: The vehicle is simply asked to travel straight ahead, keeping itself aligned parallel to the road edges.

SINGLE-LANE OVERTAKING

For this problem, it is assumed that the road is marked in the middle by a real or virtual boundary separating the outbound and incoming directions of travel. The behaviours introduced in Section 2 are all used, and their computations are done using this boundary. As a result, while driving using any of the behaviours, the vehicle does not slip into the wrong side of the road at all, even though there may be no physical barrier separating the sides. The single-lane overtaking behaviour is modelled in addition to these behaviours. The computation of the single-lane behaviour, however, uses the actual road boundaries. Various aspects of the behaviour are detailed as follows.

Single-Lane Overtaking Initiation

Consider that the vehicle being planned R_i cannot overtake a slower vehicle in front by normal means. It may attempt to see if single-lane overtaking is possible. It is required that the vehicle is currently on the correct side of the road and not already attempting single-lane overtaking. As per the general methodology of overtaking trajectory computation, we need to firstly decide whether the overtaking will happen on the left side or right side of the vehicle ahead and then select the preferred lateral position for overtaking. Here, it is assumed that the traffic operates with a "drive on the left" rule, and hence, single-lane overtaking can only happen on the right side.

Consider that the vehicle in front is R_j with a separation of r_j available on its right (without considering the virtual boundary). Furthermore, consider that the vehicle R_i has a separation r_i on its right side. The availability of these separations may be partly on the correct (outbound) carriageway and partly on the wrong (inbound) carriageway. The most important precondition for the overtaking to be initiated is that both of these separations must be larger than the minimum separation required by R_i for it to overtake, which is its width (wid_i) and a safety distance ($separMin_i$) on both sides, totalling $wid_i + 2.separMin_i$. The notations are shown in Figure 2a.

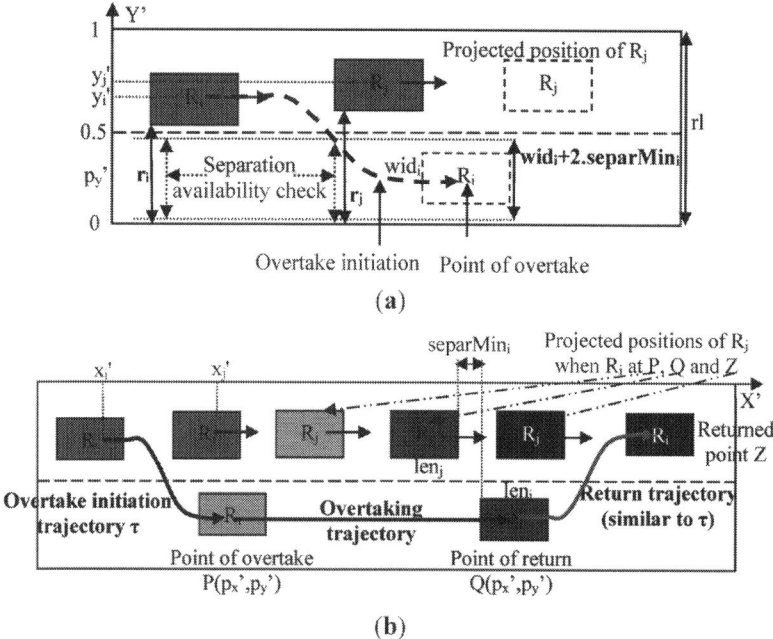

Figure 2. Notations used in single-lane overtaking. (a): For overtaking initiation; (b): Completion of overtaking.

On availability of the separation, the next task is computing a point $P(p_x',p_y')$ to which the vehicle may travel for overtaking to occur. Usually, this overtaking point would lie on the wrong side of the road. On reaching $P(p_x',p_y')$, the vehicle may travel almost straight ahead to move in front of the vehicle being overtaken and subsequently may attempt to return to the original lane. p_y' denotes the lateral position that the vehicle attempts to achieve during the overtaking procedure. The basic requirement is separation maximization, and hence, this point must be far enough from the vehicle being overtaken, as well as from any other vehicle or road boundary that may lie towards the right. The maximization is under a threshold of $separMax_i$ which disallows the vehicle from going too far on a wide road that would require large steering movements. Computation of p_y' is given by Equation (2).

$$p_y' = \begin{cases} (y_j'.rl - wid_j/2 - separMax_i - wid_i/2)/rl & r_j \geq wid_i + 2separMax_i \quad (i) \\ (y_j'.rl - wid_j/2 - r_j/2)/rl & r_j < wid_i + 2separMax_i \quad (ii) \end{cases}$$

$$(2)$$

Here, rl is the current width of the road and y_j' is the position of R_j in the Y' axis. Equation 2(i) denotes the condition when the separation available on the right of R_j is wide enough for the vehicle R_i to enjoy a maximum separation of $separMax_i$ on both sides. Equation 2(ii) denotes the condition when the largest possible separation is not available, and hence, R_i simply attempts to drive in between the separation available.

The value of p_x' is the same as per the general guidelines used in [12], given by Equation (3). The basic idea is that the curve generated for the travel of the vehicle till the point of overtake $P(p_x',p_y')$ must be smooth enough for navigation with the current speed v_i.

$$p_x' = c_1 + c_2.v_i + c_3. abs(y_i' - p_y').rl$$

$$(3)$$

Here, c_1, c_2, c_3 are constants, $abs()$ is the absolute value function and y_i' is the current position of R_i in the Y' axis. c_1 is the minimum distance on the X' axis needed to produce a smooth curve as per spline curves. Cubic splines are used with the constraint that the spline must start from the current position, headed in the current orientation of the vehicle, and should end at the point of overtake headed parallel to the road. Usually, this is set to be twice the length of the vehicle. c_2 and c_3 meanwhile denote the contributions of the factors of speed and steering requirements. Spline curves are used to compute the overtaking initiation trajectory τ from the current vehicle position R_i to $P(p_x',p_y')$, such that the vehicle is parallel to the road at $P(p_x',p_y')$.

The time required for the vehicle to place itself in the overtaking zone is given by $\|\tau\|/v_i$, where $\|\tau\|$ is the length of the trajectory and v_i is the current vehicle speed. The vehicle may then be seen to accelerate from its original speed v_i to the maximum speed $vMax_i$ in order to be well ahead of R_j (with an additional safety distance of $separMin_i$) before its return to the correct side can take place. In normal circumstances, the trajectory for the vehicle to return to the correct side of the road is similar (flipped) to the overtaking trajectory. However, its speed will have increased to $vMax_i$, giving a return time of $\|\tau\|/vMax_i$, which will be shorter than $\|\tau\|/v_i$. Overall, we can then find the total overtaking period with the corresponding expected straight trajectory, called the overtaking trajectory. The notations are shown in Figure 2b. The total time for overtaking may hence be approximated by Equation (4).

$$T = \frac{\|\tau\|}{v_i} + \frac{x_j' - x +_i' \max(len_i, len_j)/2 + separMin_i}{\frac{v_i + vMax_i}{2} - v_j} + \frac{\|\tau\|}{vMax_i}$$

(4)

Here, x_i' and x_j' denote the current position of R_i and R_j, respectively, while len_i and len_j denote the corresponding lengths. v_j is the current speed of R_j.

For single-lane overtaking to be initiated, the preconditions of the general overtaking behaviour (Section 2) are included, that is the lane change behaviour (Section 2) for trajectory τ must be true, as well as the condition that R_j must not be drifting rightward. Further, no collisions should occur between the vehicle R_i and any other vehicle (as per their projected travel) on the wrong side of the road, if R_i occupies the computed lateral position for the time interval T. This means, ideally, if the vehicles stick to their current speeds and lateral positions, the vehicle R_i should be able to complete the overtaking procedure within time T without causing any vehicles to slow down.

General Travel

In case a vehicle is travelling on the wrong side of the road and wishes to attempt single-lane overtaking, it must always test whether such overtaking is feasible or whether it has been completed, which are modelled as separate behaviours and discussed in Sections 4.3 and Sections 4.4. In all other cases, the vehicle must travel in the wrong lane using the same behaviour set as any general vehicle (travelling on the correct side of the road), with the exception that no overtaking can be performed. This would mean that mostly either the vehicle travels straight ahead or adjusts its lateral position to maximize its separation from all other vehicles.

The speed needs to be set for every vehicle in the planning scenario whilst it travels during single-lane overtaking or any other behaviour. As per the heuristics used, each vehicle must always attempt to travel within its maximum safe speed. This notion is, however, different for the different classes of vehicles possible.

The first type of vehicle (i) is a normal vehicle, which is on its correct side of travel, and no other vehicle attempting single-lane overtaking is to be found ahead (Figure 3a). For such a vehicle, it is assumed that in the worst case, the vehicle driving in front might brake suddenly, and hence, the speed should correct to enable the vehicle to react instantaneously and avoid a collision with a safety distance of *separMin$_i$*. The preferred speed is given by Equation (7(i)).

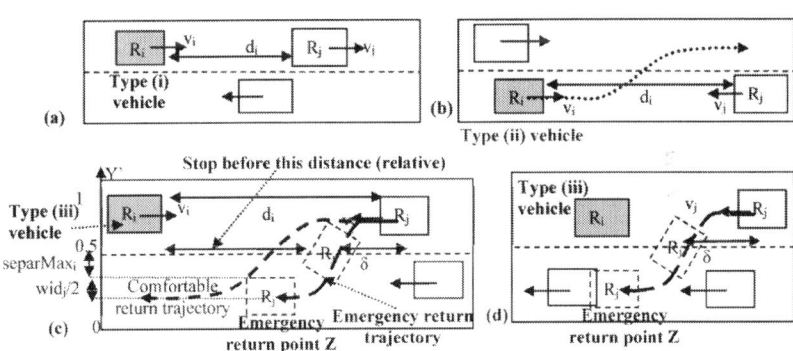

Figure 3. Various types of vehicles and notations for their speed computation.

The second category of vehicle (ii) is one which is on the wrong side of the road in the middle of attempting single-lane overtaking and finds a vehicle R_j directly ahead of it (Figure 3b). Hence, R_i and R_j are travelling towards each other with speeds v_i and v_j. The attempt here is to stop R_i, which is the vehicle being planned, before a possible collision. The speed must always be such to enable R_i to stop with the maximum possible deceleration. The speed setting is hence the same as for a normal vehicle, with the resultant safe speed interpreted as the resultant safe relative speed. The preferred speed is given by Equation (7(ii)).

The last category (iii) is if the vehicle R_i is on the correct side of the road with vehicle R_j travelling towards it on the wrong side, attempting single-lane overtaking (Figure 3c). Vehicle R_i is aware of the fact that R_j needs to travel back to its correct side, and hence, mere collision prevention is not enough. If R_i and R_j stand almost touching each other, neither can move until one of the vehicles backs up. R_j needs some distance to return to its correct side. In an extreme case, R_j should have additional space available to steer left and return to its correct lane (Figure 3d) using an emergency return trajectory, while still avoiding collision with R_i. R_j would need to stop and wait for the other vehicles to clear on the correct side and slowly trace the emergency return trajectory. In extreme cases, both the vehicles R_i and R_j should slow down sufficiently for R_j to return. This additional distance was not maintained by the earlier category (or R_j itself does not contribute much in slowing down), as it is important for R_j to get close to R_i while tracing the emergency return trajectory. R_j must return to the correct side and not stop because of the presence of R_i. Otherwise, both vehicles may wait indefinitely.

As perceived by R_i, if the vehicle R_j immediately needs to return to its original lane, it would need to choose an emergency point of return $Z(z_x', z_y')$. Here, z_y' is the chosen lateral position to which R_j would return. Assuming sufficient distance would be available with R_j, its lateral position at the return can be given by Equation (5). Y' = 0.5 is the virtual boundary from which the maximum separation distance is desired.

$$z_y' = 0.5 + (wid_j/2 + separMax_i)/rl \qquad (5)$$

The distance required (δ) along the road to change the lateral position as computed in Equation (5) is given by Equation (6), whose notations are the same as those in Equation (3). It is assumed here that half of this distance would be required on the wrong side of the road, while the other half would be on the correct side of the road, and hence, only the first half is in question. This is shown clearly in Figure 4.

$$\delta = (c_1 + c_2.v_i + c_3. Abs(y' - z_y').rl)/2 \qquad (6)$$

The vehicle R_i attempts to change its speed such that this distance can be assured even if R_i has to stop to make this distance available. The relative speed is computed, which is used to compute the actual speed given by Equation (7(iii)).

The preferred speeds for various cases are given by Equation (7), while the actual speeds considering acceleration limits are given by Equation (8). Here, d_i is the distance of R_i from the vehicle or obstacle in front.

$$vm = \begin{cases} \sqrt{2.accMax_i.\max(d_i - separMin_i, 0)} & (i) \\ \max\left(\sqrt{2.accMax_i.\max(d_i - separMin_i, 0)} - v_j, 0\right) & (ii) \\ \max\left(\sqrt{2.accMax_i.\max(d_i - separMin_i - \delta, 0)} - v_j, 0\right) & (iii) \end{cases}$$

(7)

$$v_i' = \begin{cases} \min(v_i + accMax_i, v_i^{\max}) & v_i + accMax_i \leq vm \\ \max(v_i - accMax_i, vm) & v_i > vm \\ vm & otherwise \end{cases}$$

(8)

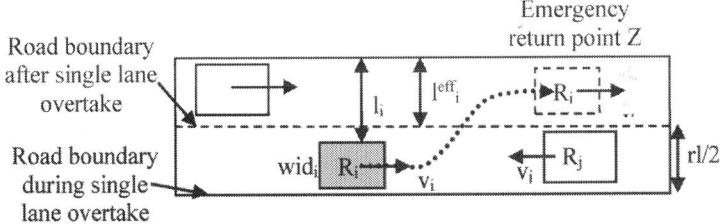

Figure 4. Return point calculation.

Cancelling Single-Lane Overtaking

While a vehicle R_i is attempting single-lane overtaking, factors unaccounted for in its initiation may stop the overtaking from being successfully completed, and the vehicle may need to return to its original side as quickly as possible. Driving on the wrong side is dangerous and creates problem for the entire traffic system. Hence, the feasibility of overtaking must be continuously tracked during the whole overtaking process. At any time, overtaking may be regarded as infeasible if there is some vehicle R_j in front, driving towards R_i and the distance between the two vehicles is just large enough (or below) the minimal distance required by R_i to execute the emergency return trajectory. This minimum distance (δ) is given by Equation (6). If the current distance is just equal or less than the required amount or R_i cannot decelerate fast enough to stop the distance (δ) from reducing below this amount, single-lane overtaking is regarded as being cancelled on that occasion.

The vehicle now needs to actually return to its original side of travel. In this stage, the actual point of emergency return $Z(z_x', z_y')$ is computed by an analysis of the vehicles on the correct side. Suppose l_i is the separation to the left that is currently available. This distance does not account for the fact that after R_i returns to its lane, the virtual boundary would become the boundary to keep away from, and hence, the effective distance available to the left is that beyond the virtual boundary given by Equation (9). The notations are shown in Figure 4.

$$l^{eff}_i = l_i - (rl/2 - y'_i.rl - wid_j/2) \tag{9}$$

Based on the guidelines of separation maximization under a threshold of $separMax_i$, the desired lateral position of the vehicle is given by Equation (10). In case the minimum separation is not available, R_i needs to attempt to return back to the correct side whilst maintaining at least a minimum separation.

$$
Z'_y =
\begin{cases}
0.5 + (wid_j/2 + separMin_i)/rl & l^{eff}_i \leq wid_i + 2.separMin_i \\
0.5 + (wid_j/2 + separMax_i)/rl & l^{eff}_i \geq wid_i + 2.separMax_i \\
0.5 + l^{eff}_i/2.rl & wid_i + 2.separMax_i < l^{eff}_i < wid_i + 2.separMax_i
\end{cases}
\tag{10}
$$

The (emergency) return trajectory τ is computed using spline curves joining the current point to the point of return Z. The trajectory is traversed at all times with the highest possible speeds as calculated in Equation (8). Since the return trajectory might not be clear, it is evident that the vehicle may have to slow down drastically and even stop and wait for some vehicles to clear. However, it would eventually return to its original side.

Completing Single-Lane Overtaking

Once R_i, which was attempting single-lane overtaking, is ahead of the vehicle being overtaken, it may aim to return to its original side of travel. This behaviour is in fact the same as cancelling the single-lane overtaking behaviour with the exception that it is invoked post overtaking and only when the minimum separation is available. In the absence of minimum separation, the vehicle continues to traverse on the wrong side of the road. It may return when the minimum separation is available or may be forced to return by cancellation of the single-lane overtaking behaviour when a vehicle in front, travelling in the opposite direction, is too near. Implementation details are then the same as for the cancelling single-lane overtaking behaviour.

The resultant set of behaviours is summarized in Table 1, and the resultant algorithm is given as Algorithm 1. Table 1 and Algorithm 1 are extended from [12].

Algorithm 1:

Plan (Vehicle R_i, Map, Previous plan τ)

If new obstacle found
- Compute τ for obstacle avoidance
- If lane_change (τ), return τ
- Else, $v_i \leftarrow \max (v_i - accMax_i, 0)$, $R_i \leftarrow$ move a unit step, return null

If no slower vehicle and obstacle ahead in vicinity $\wedge$ v_i close to v_i^{max} $\wedge$ $\tau =$ null
- $CEN \leftarrow (x_i' + \Delta(v_i), 0.5)$
- $\tau \leftarrow$ curve (R_i, CEN)
- If $\tau(t) \in \zeta_i^{free}$ $\forall$ t, return τ

If $\tau \neq$ null
- $v_i \leftarrow$ Safe speed as per Equation (8)
- If Conflicting Interests $\wedge$ τ is non-straight
- $\tau \leftarrow$ straighten(τ), return τ
- Else If performing single-lane overtake
- Calculate δ using Equation (6)
- If distance δ not available or not expected to be available in the

future even with largest deceleration
- Compute Z for return using Equation (10)
- $\tau \leftarrow$ curve (R_i, Z)
- return τ

Else
- $R_i \leftarrow$ Move a unit step by τ
- If τ is over, return null, else return τ

Else If performing single-lane overtake
- Calculate δ using Equation (6)
- If distance δ not available or not expected to be available in the future even with largest deceleration
- Compute Z for return using Equation (10)
- $\tau \leftarrow$ curve(R_i, Z)
- return τ

If slow vehicle in front $\land$ sufficient separation exists for overtaking assuming cooperation $\land$ not undergoing single-lane overtake
- $R_j \leftarrow$ Vehicle to overtake, side $\leftarrow$ side of overtaking
- Compute P for overtake
- $\tau \leftarrow$ curve(R_i, P)
- If $\tau(t) \in \zeta_i^{free} \; \forall \; t \land$ lane_change$(\tau) \land R_j$ not steering towards side,
- return τ

If Slower vehicle in front $\land$ sufficient separation available $\land$ no collisions with vehicle in wrong side for expected time of completion of overtake assuming no cooperation
- $R_j \leftarrow$ Vehicle to overtake
- Compute P for single-lane overtake using Equations (2) and (3)
- $\tau \leftarrow$ curve(R_i, P)
- Calculate T by Equation (4)
- If $\tau(t) \in \zeta_i^{free} \; \forall \; t \land$ lane_change$(\tau) \land$ no collisions with vehicles in wrong side till T assuming no cooperation $\land R_j$ not steering towards side
- return τ

If performing single-lane overtake $\land$ sufficient separation available at correct side
- Compute Z for return using Equation (10)

- $\tau \leftarrow$ curve(R_i,Z)
- If $\tau(t) \in \zeta_i^{free} \ \forall \ t \wedge$ lane_change(τ), return τ

If vehicle overtaking at back $\wedge$ separation available to offer $\wedge$ not undergoing single-lane overtake
- $R_j \leftarrow$ Vehicle to allow overtake, side $\leftarrow$ side of being overtaking
- Compute P for being overtaken
- $\tau \leftarrow$ curve(R_i,P)
- If $\tau(t) \in \zeta_i^{free} \ \forall \ t \wedge$ lane_change(τ), return τ

If $l_i + r_i \geq 2.\text{min\_separ}_i \wedge (l_i < \text{max\_separ}_i \wedge r_i < \text{max\_separ}_i)$
- Compute P for separation maintenance
- $\tau \leftarrow$ curve(R_i,P)
- If $\tau(t) \in \zeta_i^{free} \ \forall \ t \wedge$ lane_change(τ), return τ

If $l_i + r_i < 2.\text{min\_separ}_i$
- $v_i \leftarrow \max(v_i - \text{accMax}_i, 0)$, $R_i \leftarrow$ move a unit step, return null
- $v_i \leftarrow$ Safe speed as per Equation (8)
- $R_i \leftarrow$ Move a unit step parallel to the road
- return null

Table 1. Summary of vehicle behaviours.

S. No.	Behaviour	Pre-Condition	Description	In-Behaviour Specifications	Priority
1.	Obstacle avoidance	Obstacle discovery, lane change true	Strategy to avoid obstacle	Check for collisions with vehicle in front, obstacle avoidance	1
2.	Centring	No vehicle, no obstacle, vehicle travelling at high speed	Put vehicle in the centre of the road	NIL	2
3.	Lane	Called by	Whether	NIL	NA

	change	other behaviours	possible to steer		
4.	Overtake	Slower vehicle ahead, sufficient separation available assuming the cooperation of all vehicles ahead, lane change true, not undergoing single-lane overtaking	Strategy to initiate overtaking, ask other vehicles to move and eventually align, so that travelling straight completes overtaking	Discover conflicting interests, check for collisions with vehicles in front	3
5.	Single-lane overtaking	Slower vehicle ahead, sufficient separation available, lane change true, no collisions with a vehicle on the wrong side for the expected time of completion of overtaking assuming	Attempt to initiate single-lane overtaking by placing the vehicle on the wrong side	Check for cancellation of single-lane overtaking, check for collision with vehicle in front	4

		no cooperation			
6.	Cancel single-lane overtaking	Performing single-lane overtaking, sufficient separation not available or not expected to be available in the future even with the largest deceleration to return to the original lane because of a vehicle ahead on the wrong side	Attempt to place the vehicle on the correct side and not allow any subsequent motion on the wrong side	Check for collision with the vehicle in front and if clear move ahead	NA
7.	Complete single lane overtaking	Performing single-lane overtaking, sufficient separation available on the correct side, lane change true	After completing overtaking, return to the correct side	Check for collision with the vehicle in front	5
8.	Be overtaken	Vehicle to the rear shows the need for overtaking,	Align so that the vehicle to the rear needing to	Discover conflicting interests, check for collisions	6

		separation available to offer, lane change true, not undergoing single-lane overtaking	overtake has more overtaking separation	with the vehicle in front	
9.	Maintain separation steer	Maximum separation possible not available at both ends, while steering in some manner can increase current lowest separation, lane change true	Steer to maintain as high separation as possible (not more than threshold) from both ends	Discover conflicting interests, check for collision with vehicle in front	7
10.	Slow down	No adjustment of steering capable of generating minimal separation at both ends	Reduce speed	NIL	8
11.	Discover conflicting interests	A neighbouring vehicle steering towards vehicle being	Straighten trajectory being followed	Check for collision with the vehicle in front	NA

		planned found too close while the vehicle being planned was steering towards it, vehicle following a non-straight trajectory			
12.	Travel straight	No pre-conditions	Take a unit step forward as per the road's current orientation	Check for collisions with the vehicle in front	9

RESULTS

The designed set of behaviours was studied via a series of simulations. The road considered was wide enough for only two vehicles to be accommodated comfortably. Half the width was reserved for traffic travelling in one (outbound) direction and the other half for traffic travelling in the other (inbound) direction. This is the most typical and the only likely scenario where such overtaking would take place. Indeed, wider roads may not necessitate that traffic from both directions mixes with each other. Different scenarios involving a single-lane overtaking were tested.

Being a very risky behaviour, most of the time, single-lane overtaking would not be attempted. Hence while we looked at many scenarios with numerous vehicles, most of them showcased no single-lane overtaking behaviour. The performance of vehicles without this behaviour was the objective of our previous work [12] and is hence not repeated in the current work. This can be practically seen in everyday life, as such overtaking is rarely seen and, then, only too when the conditions are the most favourable or idealistic. The visual display of all single-lane overtaking behaviours was the same, and hence, those results are not displayed. We tried to put the most interesting or the closest overtaking cases in the paper, as the other results are much easier to obtain.

In the first scenario, a slower vehicle (A) was generated in the centre of the road, and this vehicle naturally simply drifted towards the correct side of travel. A faster vehicle (B) was generated later. The faster vehicle (B) judged the presence of the slower vehicle (A) ahead and the feasibility of single-lane overtaking. By the projected motion of B, it was clear that single-lane overtaking could easily happen by the vehicle going to the wrong side, and hence, the vehicle decided to perform single-lane overtaking. B jumps to the wrong lane or overtaking lane by the initiation of the single-lane overtaking behaviour, surpasses A by general travel during the single-lane overtaking behaviour and then decides to return back to the original lane by the completion of the single-lane overtaking behaviour. There was no change at all in the motion of A, and overtaking was easily completed as shown in Figure 5a and Video in supplementary information.

The second scenario was created to test the ability of a vehicle when such overtaking is not possible. In this scenario, an additional vehicle (C) was made to appear on the other side, travelling in the opposite direction. When the faster vehicle (B) emerged, it computed the feasibility of overtaking the slower vehicle (A) by considering the motions of A and C. C made single-lane overtaking seem risky, and hence, B did not attempt to overtake A. Instead, it followed A. When C had passed both A and B, again, the feasibility of single-lane overtaking was judged. This time, there was no vehicle that could make overtaking seem infeasible or risky. Single-lane overtaking by B was now judged to be feasible. It was initiated and subsequently completed in a manner similar to Scenario 1. The trajectories of the three vehicles and the scenario are shown in Figure 5b and Video in supplementary information.

In the third scenario, vehicle B judged single-lane overtaking to be feasible and initiated the same. There was, however, an oncoming vehicle (C) travelling on the other side of the road. The emergence and speed of *C* was just enough for overtaking to be feasible and just possible without cooperation. A slower speed for *C* would have made overtaking comfortable and, thus, not challenging, whereas a higher speed would have made overtaking infeasible. By fixing the speed of *C* in such a manner, the narrow overtaking behaviour was studied. On committing to overtaking, *B* quickly placed itself on the wrong side of the road, went ahead of the vehicle being overtaken (*A*) and returned in front of *A* in the correct lane. However, *C* was driving towards *B* during single-lane overtaking, which could have been a threat. *C* assessed a possible threat in advance by computing the relative speeds of the vehicles as a part of its normal driving behaviour and showed cooperation to some extent by slowing down. This gave *A* some extra space to return back to the correct side of the road and make overtaking comfortable and risk-free. The results are shown in Figure 5c and Video in supplementary information.

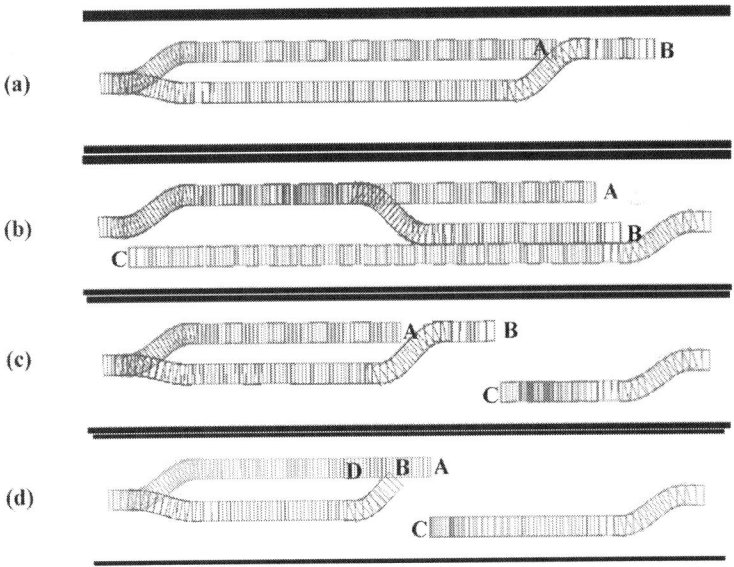

Figure 5. Experimental results. (**a**): Scenario 1—Simple single lane overtaking; (**b**): Scenario 2—Single lane overtaking with an oncoming vehicle; (**c**): Scenario 3—Single lane overtaking requiring the oncoming vehicle to slow down; (**d**): Scenario 4—Cancelling of single lane overtaking due to infeasibility by an oncoming vehicle.

The aim in the last scenario was to put the overtaking vehicle (*B*) in an awkward scenario, which is the worst that such a vehicle can enter. The vehicle was made to initiate overtaking. In doing so, it did not account for *A* ahead, which would not allow *B* to return back to the correct lane easily on completion of single-lane overtaking. Overtaking was initiated, and the vehicle did go ahead of the slower vehicle *D*, with *C* coming from the other direction. After a little general driving on the wrong side of the road, it was clear that overtaking was not complete, because of the presence of *A*, and general driving could not continue, because *C* was approaching on the other side of the road. Hence, the infeasibility of overtaking was discovered, and an emergency return trajectory was computed. However, the emergency return trajectory was infeasible, due to a possible collision with *A*. *C* detected the infeasibility of *B* during its general driving speed assignment and slowed down to give enough space for *B* to eventually return back to its original side. *B* therefore placed itself in between the two slower vehicles (*D* and *A*). *D* saw *B* trying to accommodate during its general driving, and when *B* was able to slip in, *D* did well to give *B* accommodated in between. The scenarios are shown in Figure 5d and Video in supplementary information.

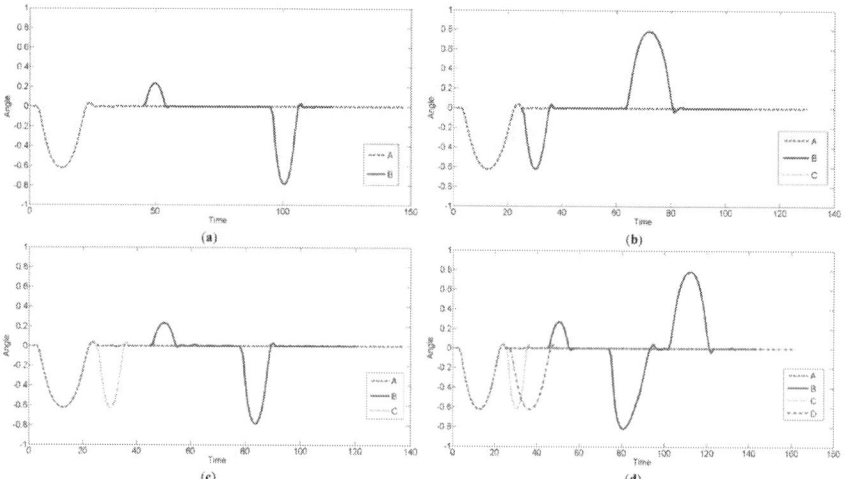

Figure 6. Angle profiles of different vehicles for different scenarios. (**a**): Scenario 1; (**b**): Scenario 2; (**c**): Scenario 3; (**d**): Scenario 4.

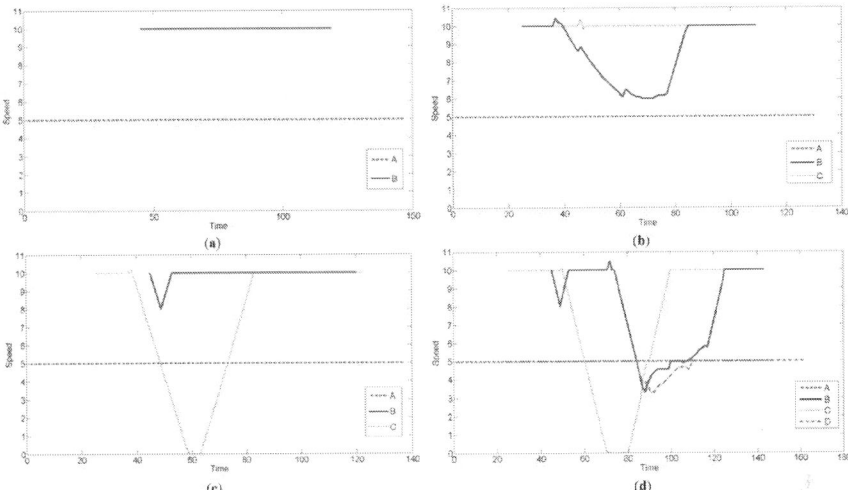

Figure 7. Speed profiles of different vehicles for different scenarios. (**a**): Scenario 1; (**b**): Scenario 2; (**c**): Scenario 3; (**d**): Scenario 4.

Table 2. Summarization of the results.

Scenario	Vehicle	Time to Destination	Distance Travelled till Termination	Maximum Allowed Speed	Average Speed
1.	A	147	716.2713	5	4.8726
	B	75	726.2893	10	9.6839
2.	A	130	632.2713	5	4.8636
	B	85	690.4683	10	8.1232
	C	74	716.1901	10	9.6782
3.	A	137	667.8519	5	4.8748
	B	76	730.2858	10	9.6090
	C	99	721.2544	10	7.2854
4.	A	147	716.2713	5	4.8726
	B	99	746.8892	10	7.5443
	C	103	714.8350	10	6.9401
	D	138	645.2713	5	4.6759

The statistical results relating to different scenarios are given in Table 2. Further, the angle and speed profiles for different vehicles are given in Figure 6 and Figure 7. In Figure 6a, both B and C turn towards their left simultaneously on entering the scenario, using similar dynamics. Hence, their angle profiles intersect. The units of distance and time are arbitrary and depend on the simulator. These can be mapped to real-world units by appropriate constants.

CONCLUSIONS

Enabling autonomous vehicles to drive in unorganized and chaotic traffic is a difficult problem that requires due consideration from a research perspective. In this paper, we made a significant step while considering traffic planning on straight roads as the specific problem being attacked. The work extended our earlier planning framework in which the assumption was that all vehicles were merely travelling on the same side of the road. By using the behaviours presented and experimented on in this paper, the framework can now be used in a variety of scenarios, both where roads are separated with a physical barrier or are just one-way and when the roads operate as a dual carriageway with no physical barrier. In particular, this paper looked in depth at the important overtaking procedure wherein a vehicle for some time occupies a part of the wrong side of the road in order to accomplish its overtaking activity. It is worth remembering that if such overtaking were not allowed, then the overall performance of the system would be much lower, with vehicles reaching their destination at a later time due to travelling for longer distances behind slower vehicles.

In the paper, we looked at aspects of assessing the feasibility of such overtaking and also at formulating an initiation trajectory, the mechanisms involved with driving on the wrong side, judging the infeasibility of such overtaking and cancelling and successfully completing an overtaking procedure. This behaviour was used in priority with the behaviours proposed earlier and practically meant that if normal overtaking, on the same side of the road, is not possible, then a vehicle would check for the possibility of single-lane overtaking.

Experimental evaluations were carried out by means of simulations. In particular, the single-lane overtaking behaviour was tested in various scenarios. In all cases, the vehicle either successfully completed the overtaking action or was able to cancel doing the same. No collisions were recorded using our method, and the vehicles apparently behaved in a similar fashion to actual vehicles on the road performing such overtaking. Currently, it is unfortunately neither possible to employ the algorithm on an actual vehicle and test it in real traffic, nor is it possible to verify such behaviour against standard traffic datasets. All traffic datasets are for organized traffic, while the work described here is for unorganized traffic. The difficulty of tracking vehicles in the absence of speed lanes makes obtaining such a dataset difficult. Having a dataset recording the motion of vehicles in real chaotic traffic would though enable the verification of the algorithm described here, as well as help to mine for new behaviours that could be used for planning. An autonomous vehicle requires different modules apart from planning, which need to be able to work in the absence of speed lanes. Clearly, these aspects can all be considered as future research.

ACKNOWLEDGMENTS

The first author was supported by the Commonwealth Scholarship Commission in the United Kingdom and the British Council under Commonwealth Scholarship and Fellowship Program (2010), U.K., Award Number INCS-2010-161.

AUTHOR CONTRIBUTIONS

R. K. and K. W. conceived and designed the algorithms; R. K. developed and tested the algorithms; R. K. and K. W. wrote the paper.

REFERENCES

1. Buehler, M.; Iagnemma, K.; Singh, S. *The 2005 DARPA Grand Challenge: The Great Robot Race*; Springer-Verlag: Berlin/Heidelberg, Germany, 2007.
2. Montemerlo, M.; Becker, J.; Bhat, S.; Dahlkamp, H.; Dolgov, D.; Ettinger, S.; Haehnel, D.; Hilden, T.; Hoffmann, G.; Huhnke, B.; *et al*. Junior: The Stanford entry in the Urban Challenge. *J. Field Robot.* 2008, *25*, 569–597.
3. Bishop, R. Intelligent vehicle applications worldwide. *IEEE Intell. Syst. Their Appl.* 2000, *15*, 78–81.
4. Bishop, R. *Intelligent Vehicle Technology and Trends*; Artech House Publishers: Norwood, MA, USA, 2005.
5. Kasper, D.; Weidl, G.; Dang, T.; Breuel, G.; Tamke, A.; Wedel, A.; Rosenstiel, W. Object-Oriented Bayesian Networks for Detection of Lane Change Maneuvers. *IEEE Intell. Transp. Syst. Mag.* 2012, *4*, 19–31.

6. Reveliotis, S.A.; Roszkowska, E. Conflict Resolution in Free-Ranging Multivehicle Systems: A Resource Allocation Paradigm. *IEEE Trans. Robot.* 2011, *27*, 283–296.

7. Urmson, C.; Baker, C.; Dolan, J.M.; Rybski, P.; Salesky, B.; Whittaker, W.L.; Ferguson, D.; Darms, M. Autonomous driving in traffic: Boss and the urban challenge. *AI Mag.* 2009, *30*, 17–29.

8. Paruchuri, P.; Pullalarevu, A.R.; Karlapalem, K. Multiagent simulation of unorganized traffic. In Proceedings of the First International Joint Conference on Autonomous Agents and Multiagent Systems: Part1, Bologna, Italy, 15–19 July 2002; ACM: New York, NY, USA, 2002; pp. 176–183.

9. Vanajakshi, L.; Subramanian, S.C.; Sivanandan, R. Travel time prediction under heterogeneous traffic conditions using global positioning system data from buses. *IET Intell. Transp. Syst.* 2009,*3*, 1–9.

10. Mohan, D. Traffic safety and health in Indian cities. *J. Transp. Infrastruct.* 2002, *9*, 79–94.

11. Jain, A.; Menezes, R.G.; Kanchan, T.; Gagan, S.; Jain, R. Two wheeler accidents on Indian roads—A study from Mangalore, India. *J. Forensic Leg. Med.* 2009, *16*, 130–133.

12. Kala, R.; Warwick, K. Motion planning of autonomous vehicles in a non-autonomous vehicle environment without speed lanes. *Eng. Appl. Artif. Intell.* 2013, *26*, 1588–1601.

13. Aguirre, E.; Gonzalez, A. Fuzzy behaviours for mobile robot navigation: Design, coordination and fusion. *Int. J. Approx. Reason.* 2000, *25*, 255–289.

14. Dee, H.M.; Hogg, D.C. Navigational strategies in behaviour modelling. *Artif. Intell.* 2009, *173*, 329–342.

15. Furda, A.; Vlacic, L. Enabling safe autonomous driving in real-world city traffic using multiple criteria decision making. *IEEE Intell. Transp. Syst. Mag.* 2011, *3*, 4–17.

16. Schubert, R. Evaluating the Utility of Driving: Toward Automated Decision Making Under Uncertainty. *IEEE Trans. Intell. Transp. Syst.* 2012, *13*, 354–364.

17. Schubert, R.; Schulze, K.; Wanielik, G. Situation Assessment for Automatic Lane-Change Maneuvers. *IEEE Trans. Intell. Transp. Syst.* 2010, *11*, 607–616.

18. Naranjo, J.E.; González, C.; García, R.; de Pedro, T. Lane-Change Fuzzy Control in Autonomous Vehicles for the Overtaking Maneuver. *IEEE Trans. Intell. Transp. Syst.* 2008, *9*, 438–450.

19. Jin-ying, H.; Hong-xia, P.; Xi-wang, Y.; Jing-da, L. Fuzzy Controller Design of Autonomy Overtaking System. In Proceedings of the 12th IEEE International Conference on Intelligent Engineering Systems, Miami, FL, USA, 25–29 February 2008; IEEE: Miami, FL, USA, 2008; pp. 281–285.
20. Petrov, P.; Nashashibi, F. Modeling and Nonlinear Adaptive Control for Autonomous Vehicle Overtaking. *IEEE Trans. Intell. Transp. Syst.* 2014, *15*, 1643–1656.
21. Hegeman, G.; Tapani, A.; Hoogendoorn, S. Overtaking assistant assessment using traffic simulation. *Transp. Res. Part C: Emerg. Technol.* 2009, *17*, 617–630.
22. Wang, F.; Yang, M.; Yang, R. Conflict-Probability-Estimation-Based Overtaking for Intelligent Vehicles. *IEEE Trans. Intell. Transp. Syst.* 2009, *10*, 366–370.
23. Karaduman, O.; Eren, H.; Kurum, H.; Celenk, M. Interactive risky behavior model for 3-car overtaking scenario using joint Bayesian network. In Proceedings of the 2013 IEEE Intelligent Vehicles Symposium, Gold Coast, QLD, USA, 23–26 June 2013; IEEE: Gold Coast, QLD, USA, 2013; pp. 1279–1284.
24. Olaverri-Monreal, C.; Gomes, P.; Fernandes, R.; Vieira, F.; Ferreira, M. The See-Through System: A VANET-enabled assistant for overtaking maneuvers. In Proceedings of the 2010 IEEE Intelligent Vehicles Symposium, San Diego, CA, USA, 21–24 June 2010; IEEE: San Diego, CA, USA, 2010; pp. 123–128.
25. Milanés, V.; Llorca, D.F.; Villagrá, J.; Pérez, J.; Fernández, C.; Parra, I.; González, C.; Sotelo, M.A. Intelligent automatic overtaking system using vision for vehicle detection. *Expert Syst. Appl.* 2012, *39*, 3362–3373.
26. Kuwata, Y.; Karaman, S.; Teo, J.; Frazzoli, E.; How, J.P.; Fiore, G. Real-timemotion planning with applications to autonomous urban driving. *IEEE Trans. Control Syst. Technol.* 2009, *17*, 1105–1118.
27. Gehrig, S.K.; Stein, F.J. Collision avoidance for vehicle-following systems. *IEEE Trans. Intell. Transp. Syst.* 2007, *8*, 233–244.
28. Kala, R.; Warwick, K. Multi-level planning for semi-autonomous vehicles in traffic scenarios based on separation maximization. *J. Intell. Robot. Syst.* 2013, *72*, 559–590.
29. Frese, C.; Beyerer, J. A comparison of motion planning algorithms for cooperative collision avoidance of multiple cognitive automobiles. In Proceedings of the 2011 IEEE Intelligent Vehicles Symposium, Baden-Baden, Germany, 5–9 June 2011; IEEE: Baden-Baden, Germany, 2011; pp. 1156–1162.
30. Sewall, J.; van den Berg, J.; Lin, M.C.; Manocha, D. Virtualized Traffic: Reconstructing Traffic Flows from Discrete Spatio-temporal Data. *IEEE Trans. Vis. Comput. Graph.* 2011, *17*, 26–37. [PubMed]

31. Chu, K.; Lee, M.; Sunwoo, M. Local Path Planning for Off-Road Autonomous Driving With Avoidance of Static Obstacles. *IEEE Trans. Intell. Transp. Syst.* 2012, *13*, 1599–1616.

32. Anderson, S.J.; Karumanchi, S.B.; Iagnemma, K. Constraint-based planning and control for safe, semi-autonomous operation of vehicles. In Proceedings of the 2012 IEEE Intelligent Vehicles Symposium, Alcala de Henares, Spain, 3–7 June 2012; IEEE: Madrid, Spain, 2012; pp. 383–388.

33. Kala, R.; Warwick, K. Planning Autonomous Vehicles in the Absence of Speed Lanes using an Elastic Strip. *IEEE Trans. Intell. Transp. Syst.* 2013, *14*, 1743–1752.

34. Sezer, V.; Gokasan, M. A novel obstacle avoidance algorithm: Follow the gap method. *Robot. Auton. Syst.* 2012, *60*, 1123–1134.

35. Kala, R. Multi-Robot Path Planning using Co-Evolutionary Genetic Programming. *Expert Syst. Appl.* 2012, *39*, 3817–3831.

36. Ryan, M.R.K. Exploiting Subgraph Structure in Multi-Robot Path Planning. *J. Artif. Intell. Res.* 2008, *31*, 497–542.

37. Rashid, A.T.; Ali, A.A.; Frasca, M.; Fortuna, L. Multi-robot collision-free navigation based on reciprocal orientation. *Robot. Auton. Syst.* 2012, *60*, 1221–1230.

38. Daniel, K.; Nash, A.; Koenig, S.; Felner, A. Theta*: Any-Angle Path Planning on Grids. *J. Artif. Intell. Res.* 2010, *39*, 533–579.

39. Sgorbissa, A.; Zaccaria, R. Planning and obstacle avoidance in mobile robotics. *Robot. Auton. Syst.* 2012, *60*, 628–638.

40. Wu, F.; Zilbersteinb, S.; Chena, X. Online planning for multi-agent systems with bounded communication. *Artif. Intell.* 2011, *175*, 487–511.

CITATION

Rahul Kala and Kevin Warwick, Motion Planning of Autonomous Vehicles on a Dual Carriageway without Speed Lanes, doi:10.3390/electronics4010059

CHAPTER 5

Traffic Measurement on Multiple Drive Lanes with Wireless Ultrasonic Sensors

Soobin Jeon [1], Eil Kwon [2] and Inbum Jung [1],*

[1]Department of Computer Information and Communication Engineering, Kangwon National University, Chuncheon, 200-701, Korea; [2]Civil Engineering, University of Minnesota Duluth, Duluth, MN 55812, USA

ABSTRACT

An automated traffic measuring system for use on multiple drive lanes is proposed in this paper. This system, which uses ultrasonic sensors and a lateral scanning method, is suitable for use on real traffic roads. The proposed system can be easily established and maintained in various roadway environments. In addition, the system can be adjusted to measure traffic volumes according to the size and number of drive lanes. This paper describes the results of an experiment that the lateral scanning method can be easily applied to real traffic roads and provide a low error rate and real-time responses. This system can play an important role in accurately measuring traffic volumes as part of an intelligent transportation system.

INTRODUCTION

The development of intelligent transportation systems has had a great influence on many aspects of road transportation systems. Traffic measurement technology in particular, conducted using various types of detection devices, has had an effect on the analysis of traffic flow. Loop detectors, which are installed beneath road surfaces, have been mainly used in the past for traffic measurement. However, loop detectors often break as a result of damage from vehicles that pass over them, and they have high maintenance costs [1].

Ultrasonic sensors are frequently used as vehicle detection devices because they are cheaper and more accurate than other types of devices. Most ultrasonic sensors detect vehicles by measuring from top to bottom or from side to side diagonally. However, these approaches require that a detector be installed for each lane because each detector measures only one lane on a road. Furthermore, ultrasonic sensors require considerable infrastructure on a road [2].

In this paper, we propose a lateral scanning method for measuring the traffic on multiple lanes of a road using a wireless sensor network and ultrasonic sensors. The proposed system detects and classifies vehicles from the side of a road. Whenever an ultrasonic sensor detects the passage of a vehicle on the road, the system measures the distance to the corresponding vehicle based on its lane location. Vehicle detection and lane classification can be performed using the data on the distance to the vehicle and the time required for the vehicle to pass through the detection range of the ultrasonic sensor. The detection algorithm consists of three parts. The first part calculates thresholds, which are points at which vehicles are detected in each lane. The second part filters out unnecessary data, such as noise from the natural environment. The third part determines the locations of vehicles in multiple lanes and calculates the traffic volume, based on the filtered data and the calculated thresholds.

To test our proposed system, we developed a new device called a wireless ultrasonic sensor mote (WUSM), which is small and easy to install and can be adapted for use in a variety of real road environments. The accuracy and efficiency of the proposed system for vehicle detection and classification using WUSMs were evaluated. The results of detailed experiments show that the proposed system be easily established and maintained in various roadway environments and can measure traffic volumes by vehicle size and the number of drive lanes.

The remainder of this paper is structured as follows: first an overview of existing ultrasonic sensor technology is provided. The proposed vehicle detection algorithm and system module are then described. The experiments conducted to evaluate the performance of the proposed system and the results of those experiments are summarized. Finally, conclusions drawn from the results and plans for future work are presented.

OVERVIEW OF ULTRASONIC DETECTION SYSTEMS

The collection of vehicle information, including the traffic flow rate and the presence, speed, and location of vehicles, is the most basic requirement for an intelligent transportation systems (ITS). Surveillance technologies can be classified as intrusive or non-intrusive technologies. Intrusive traffic sensors are installed within or across a pavement. Non-intrusive sensors can be installed above or on the sides of roads, with minimum disruption to traffic flow.

Vehicle sensing technologies include inductive loops, which are most commonly used in roads, pneumatic road tubes, piezoelectric cables, and weigh-in-motion systems. In general, these technologies require installation directly into or onto pavements by insertion through holes or tunneling under the surface, and they are most often used to accurately detect the presence of vehicles on the pavement. However, intrusive detection technologies have drawbacks, including disruption of traffic for installation and repair, failures induced by poor road conditions, system reinstallation necessitated by road repairs or resurfacing, and high maintenance costs.

Non-intrusive technologies include microwave radar, infrared (IR)-based systems, video image processing (VIP), and ultrasonic detectors. These devices are usually set up on the roadside or overhead position. Non-intrusive technologies therefore offer advantages that include low installation costs, ease of access for maintenance, and installation and repair without disruption of traffic. However, non-intrusive detectors also have drawbacks, include their relatively low accuracy compared to that of intrusive detectors.

The main advantage of microwave radar is that the system performance is not affected by weather changes. However, microwave radar cannot detect motionless vehicles without the aid of an auxiliary device. Infrared systems are capable of transmitting multiple beams for purposes of multi-zone detection using a single detector unit, but their performance is greatly affected by the environment: sunlight can confuse the signals, and IR energy can be absorbed and scattered by atmospheric particulates, fog, rain, and snow. The performance of VIP systems is also greatly affected by inclement weather: false detections can be caused by vehicle shadows falling on adjacent lanes, and camera vibrations can be caused by strong winds. Other disadvantages of VIP systems include the height at which the cameras must be mounted above the road (up to 60 feet high), the relatively high installation and equipment costs, and the fact that the system is only cost effective if many detection zones are required within the field of view of the camera [1,3,4].

The system proposed in this paper employs a detection algorithm based on ultrasonic sensors. "Ultrasonic" refers to high-frequency sound waves that are beyond a human's audible range. Waves with frequencies between 25 and 50 kHz are commonly used. The principal mechanism is similar to that of microwave radar. Sound pulses are transmitted, the reflected pulses are received, and the distance from the receiver to the road or a vehicle surface is calculated from the wave travel time. The performance of ultrasonic sensors is much better than that of other types of pulse devices. Ultrasonic detection systems can detect vehicles in multiple zones and measure their speeds, and they are much cheaper than intrusive systems. Also, they have disadvantages that its performance is affected by temperature change and air turbulence. However, some modern models, such as those used in our proposed system, have built-in temperature compensation.

Typically, an ultrasonic sensor transmits a sound pulse from above the road and measures the reflected pulses from the vehicle or ground, as shown in Figure 1a. Once the default distance from the detector to the ground is set, if a vehicle passes through the detection range of the ultrasonic sensor, the distance value changes depending on the vehicle's size, and the detection system detects the presence of the vehicle based on the distance data received. The vehicle detection accuracy achieved using this method is approximately 99.5% for each ultrasonic sensor installed on each lane. This method has the advantage of being able to detect vehicles accurately, but it has the disadvantage that a sensor must be installed on each lane. There is also a risk of damage to the sensors, depending on the heights of the vehicles [5–7].

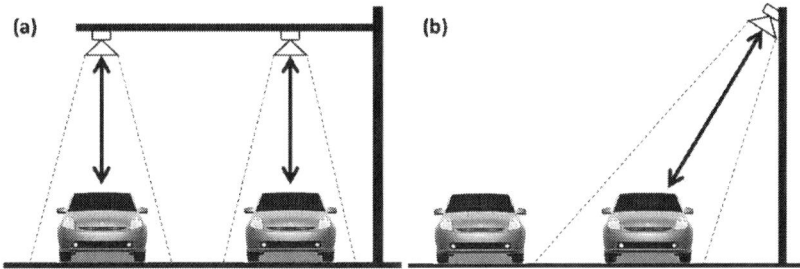

Figure 1. Ultrasonic sensor detection methods. (**a**) Overhead mount and (**b**) side top mount.

The method represented in Figure 1b detects diagonal distances to vehicles from roadside-mounted sensors. When the sensors receive waves reflected from the edges of vehicles, they detect the vehicles' presence and speeds based on the distances from the vehicles. The accuracy of this method, which is approximately 93%–95%, is lower than the previously described method by which vehicles are detected by sensors mounted above the road, because wave signals reflected only from the edges of vehicles are weak. This method also has the disadvantage that a sensor must be installed for each lane [8,9].

DEVELOPMENT OF A LATERAL SCANNING-BASED ULTRASONIC SENSING SYSTEM

Architecture

The ultrasonic sensor system in current use can only detect vehicles in one lane with one device, and it is not easy to install many devices on the side of a road because the devices are large and expensive. To address the problem of only being able to detect vehicles in one lane with one device, we propose a new detection system that also offers the advantages of easy mobility via a wireless sensor network and miniaturization via specially fabricated measuring devices. In addition, the new system can detect vehicles in multiple lanes with just one device.

To detect vehicles using ultrasonic sensors, we use a side-fire configuration with ultrasonic sensors positioned diagonally on the side of the road. Figure 2a shows the side-fire configuration for vehicle detection proposed in this paper, including the location of first lane and second lane from the center line and the location of ultrasonic sensor. The upper drawing in Figure 2a shows a vehicle in the first lane, and the lower drawing shows a vehicle in the second lane.

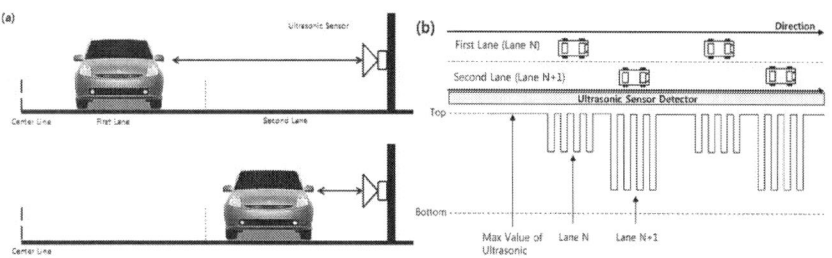

Figure 2. (a) Side-fire configuration for vehicle detection and (b) detection data from wireless ultrasonic sensor.

When an ultrasonic wave is generated, if the distance traveled by the carrier wave equals the maximum value of the sensor distance, it means there is no vehicle in the detection range. If the sensor receives carrier wave data indicating a distance less than the maximum value, it means that a vehicle is present in the lane.

Figure 2b represents schematically how measurement data received by the ultrasonic sensor are returned from vehicles on the road along a time line. In Figure 2b, the top sign of the measurement data denotes the maximum distance from the ultrasonic sensor, and the bottom sign denotes the minimum distance. Measurements that equal the maximum value indicate that there are no vehicles in the lanes. As vehicles pass the sensor zone of the ultrasonic sensor, the distance between each vehicle and the ultrasonic sensor affects the measurement data. Based on the measured lengths, the proposed system detects the locations of vehicle in multiple lanes. In Figure 2b, we see lengths labeled as corresponding lane N and lane N+1. These represent distances to the vehicles moving in lane N and lane N+1. Three types of distances are represented: the maximum value for no vehicle, the distance to a vehicle in lane N, and the distance to a vehicle in lane N+1. Based on these measurements, the proposed system can accurately detect vehicles in each lane.

Vehicle Detection Algorithm

The detection algorithm is composed of three parts. First, the "Calculate Threshold" part calculates thresholds, which are points at which vehicles are detected in each lane. Second, the "Noise filtering" part eliminates unnecessary data. Third, the "Vehicle & Lane Detection" part determines the locations of vehicles in multiple lanes and calculates the traffic volume.

Calculate Threshold

As shown in Figure 2b, when a vehicle passes the sensor zone, the ultrasonic sensor collects the carrier wave, which provides data on the distance to the vehicle. These data are used to detect the vehicles and to distinguish each lane. The distances change depending on the distances of the vehicles from the sensor. However, because there is no standard to assess the correctness of the data, it is difficult to determine the location of each vehicle.

To solve this problem, we define thresholds that represent the minimum boundary values for establishing the presence of vehicle in each lane. Figure 3a shows the patterns of distance data when Vehicle 1 and Vehicle 2 are moving in Lane N and Lane N+1. The boundary of Lane N is determined from the N threshold to 0 (zero distance). The boundary of Lane N+1 is determined from the N+1 threshold to the N threshold or max of distance. The max of distance is set according to number of lanes. For example, if the distance data for Vehicle 1 cross the threshold boundary of Lane N, it means that Vehicle 1 is moving in Lane N. Similarly, the distance data for Vehicle 2 indicate that Vehicle 2 is moving in Lane N+1.

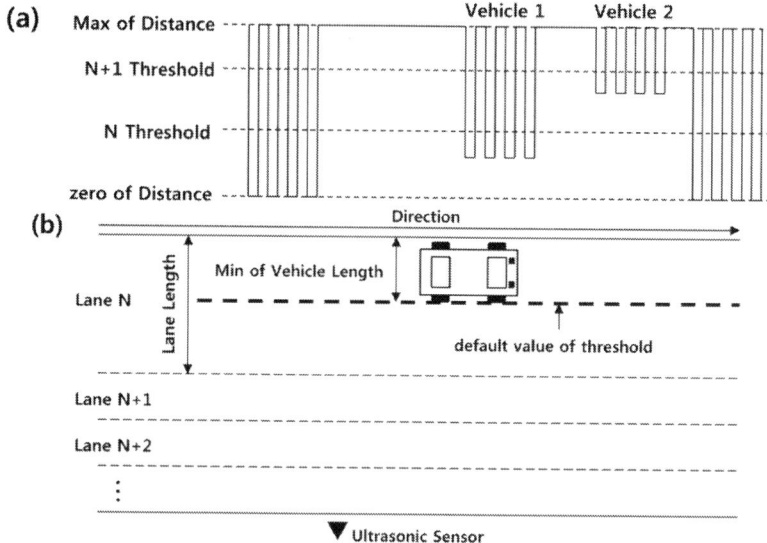

Figure 3. (a) Classification of threshold and (b) determination of default threshold.

The threshold boundaries are related to the widths of the lanes. The widths of the lanes for various types of roads should be defined according to the Department of Transportation in each country. The default value of the threshold (*DVT*) for a lane is determined based on these widths. On the other hand, the position of a sensor may change depending on the road status or environment. If the theoretical width of a lane is used without correction in determining the *DVT*, many errors may occur in vehicle detection and classification.

To address this problem, we established the *DVT* on the basis of both the minimum values of the widths of vehicles and the lane width data. The minimum values of the widths of vehicles are based on statistical data [10] that were examined for most vehicles on the road. If a *DVT* is set up to this value, it is possible to detect all vehicles on the road. Thus, assuming that *N* is Number of Lane, *WL* is the width of the lane and *MWV* is the minimum value of the width of the vehicle as shown in Figure 3b, the *DVTs* for each lane were defined as follows:

$$DVT_n = \sum_{k=1}^{N} WL \times k - MWV \qquad (1)$$

Even if an initial DVT were determined from Equation (1), it may differ depending on the actual road environment because it is a theoretical value based on statistical vehicle data and the standard lane width data. In this study, to reduce errors between theoretical data and actual road environments, we used corrected values of the thresholds ($CVTs$), with the corrections based on real field data obtained on actual roads. To calculate the CVT, an actual road vehicle is measured by threshold calculated at Equation (1). For example, if the first vehicle passes through the sensor zone and the location of the vehicle is in the DVT_1 (lane 1) boundary, the distance from the first vehicle to the sensor is CVT_1 value. The $CVTs$ were calculated based on measurements of 500–1000 vehicles (NoV) using DVT method on real roads. Equation (2) shows how the final threshold (FT) can be calculated from the lowest value of the difference between the DVT and the CVT:

$$FT_n$$
$$= MIN \left[(DVT_n - CVT_{n,k}), (DVT_n - CVT_{n,k+1}), \cdots (DVT_n - CVT_{n,k+NoV}) \right.$$

$$(2)$$

Noise Filtering

The ultrasonic sensor typically receives noise data that can lead to errors in the vehicle detection procedure. Noise data usually are generated as a vehicle approaches the sensor's angle or departs from it. When the vehicle departs from the sensor's angle, the sensor receives a small amount of noise because the surface area of the vehicle reflected by the carrier is also small [11,12].

For example, Figure 4 shows two cases of noise data received by an ultrasonic sensor. In the first case, only one vehicle passes through the sensor zone, as shown in Figure 4a-1. However, two instances of noise occur at time 4 and time 8–9. These noises suggest that three vehicles passed through the sensor zone rather than one vehicle. Figure 4b-1 shows a second case noise. There are two instances of noise at time 2 and time 8–9 that suggest that two vehicles passed through the sensor zone rather than no vehicles. Even if there are no vehicles in the sensor zone, noise can create errors in vehicle detection.

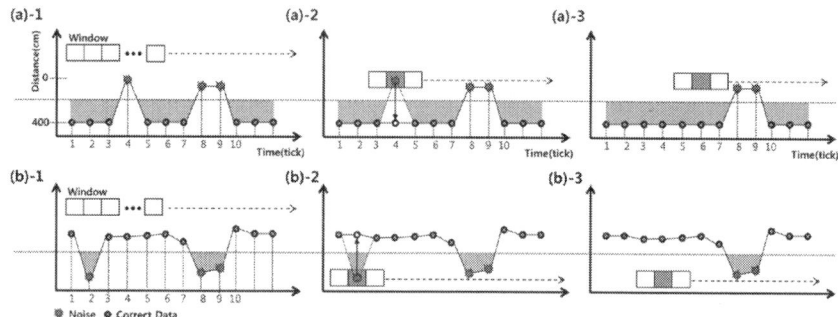

Figure 4. Noise Filtering. (**a**) Noise type-1 and (**b**) noise type-2.

The system proposed in this paper filters out noise to prevent it causing errors in detecting vehicles. To remove noise, mask windows are first installed on the noise data, as shown inFigure 4a-2,b-2. Next, because the window is applied to the data from i to $i + k$, the trend in the data is determined based on the threshold. Assuming that s is the starting point of window, k is the window size, w is the window data, ot denotes "on the threshold" and utdenotes "under the threshold", u (the trend in the noise data) is defined as follows:

$$
\begin{aligned}
u &= ot, if\ w_s = ot\ and\ w_k = ot \\
&= ut, else\ if\ w_s = ut\ and\ w_k = ut
\end{aligned}
\tag{3}
$$

If the data on both sides of window are on the threshold, the trend is ot, Otherwise, it isut. Third, assuming that NoU is the number of data points, the filtering method calculates the calibration data according to the following equation:

$$
C_i = \frac{1}{NoU} \sum_{n=s}^{k} u_n
\tag{4}
$$

The value of C_i is calculated for the window defined by $s < i < k$.

The data that do not match the trends in the window are corrected to the C_i, as shown inFigure 4a-2,b-2. Using this filtering method, noise data are removed and correct data are detected, as shown in Figure 4a-3,b-3.

In addition, only one type of noise is filtered out by three window size of mask, but there are many types of noise collected in two or more groups of vehicle data. In this study, to determine the best way to perform noise filtering, experiments were conducted using the proposed system and three to nine window size.

Detection of Vehicles and Vehicle Positions

Figure 5 represents the filtered vehicle data received as vehicles pass through the range of the sensor in each lane. Using the thresholds shown in Figure 5a, the proposed system can detect vehicles in the lanes and determine their locations.

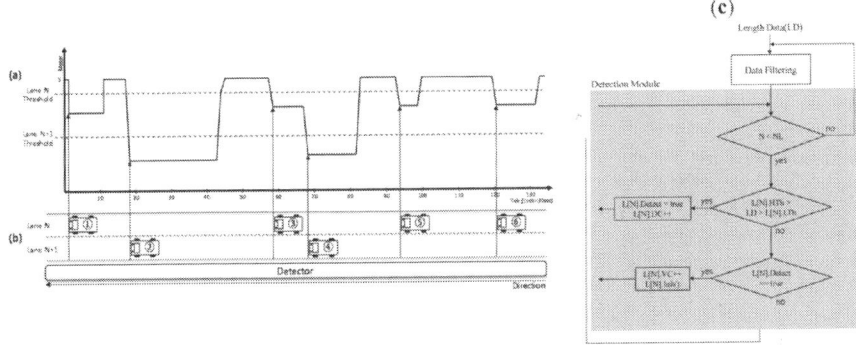

Figure 5. (a) Detected data from actual road, **(b)** location of vehicle on the road and **(c)** vehicle detection algorithm.

The presence of vehicles is determined according to the threshold of the lane. For example, if Vehicle 1 in Lane N enters the range of the sensor, data are received from the carriers of the ultrasonic sensor. The values are located between the thresholds of Lane N and Lane N+1. The horizontal lines in Figure 5a represent the range of these thresholds. While Vehicle 1 is within the range of the sensor, its data are also within the range of the Lane N threshold. After a few seconds, when the vehicle leaves the range of the sensor, the data passes over the threshold of Lane N. As a result, one vehicle is detected. Additionally, there are different patterns for detecting vehicles. Because Vehicle 3 passes under the threshold of Lane N, Vehicle 4 immediately passes under the threshold of Lane N+1. In this situation, Vehicle 3 is detected by the threshold of Lane N, and Vehicle 4 is detected by the threshold of Lane N+1.

Second, the location of each vehicle can be determined from the value of the threshold for each lane. Because they are different values, the lanes can be distinguished. For example, when a vehicle enters a sensor zone, if the sensing data are between the value of Lane N's threshold and the value of Lane N's threshold, we can determine that this vehicle is located in Lane N. Also, when there are additional lanes in the road, we can determine the vehicle location in the cases of more than Lane N according to threshold.

When there are additional lanes in the road, the proposed system can determine the vehicle location in the cases of more than Lane N because we define the thresholds that represent the minimum boundary values for establishing the presence of vehicle in each lane. The threshold boundaries are calculated with reference to widths of lanes. Eventually, the sensing distance of ultrasonic is equal to the threshold boundaries or width of lanes, and can be adjusted by number of lane. Thus, proposed system can separate the vehicle and noise as using calculated threshold in distances bigger than Lane N or more, and there will not be interference between vehicles going in opposite directions.

Additionally, When there is congested traffic on the road and vehicles in adjacent lanes do not appear to have any gaps, sensors could only detect one the vehicle, closer to the sensor and has missing errors. To address these types of problems, an additional sensors could be installed on the other side of the road. If sensors are on both sides of a road, the lanes are divided by 2, and each divided lane is included in one of both sensors. Sensing distance of each sensor is adjusted by number of lanes which is included in sensor. Also, when the sensors are installed on both sides of a road that contains the odd-lanes as a three-lanes or more, one of both sensor should contain the remaining one-lane after divided by 2.

Figure 5c shows a flow chart of the vehicle detection algorithm described above. First, the data are filtered by the "Data Filtering" module to remove noise. Second, in the detection module, the processing is repeated according to number of lanes (NL). For example, for a two-lane road, if the vehicle enters under the threshold of Lane 1 (L[1].HTh > LD > L[1].LTh, HTh: High Threshold, LTh: Low Threshold), the detection mode (L[1].Detect) of the current data is set to "true." If the vehicle escapes from the threshold of Lane 1, the vehicle count of Lane 1, (L[1].VC), is incremented by 1. "L[1].Detect" is re-set to "false" again to prepare for the next detection processing cycle. After the processing for Lane 1 is finished, this algorithm executes the detection process for the vehicle in Lane 2. The data structure for Lane 2 (L[2]) is used in the next loop of processing.

Wireless Ultrasonic Sensor Mote (WUSM)

In this study, we developed an ultrasonic sensor module that is easy to establish and maintain and is highly mobile, because of the miniaturization of the module. Figure 6 shows the structure of the ultrasonic sensor and communication module. Based on the architecture of the model, we developed the wireless ultrasonic sensor node.

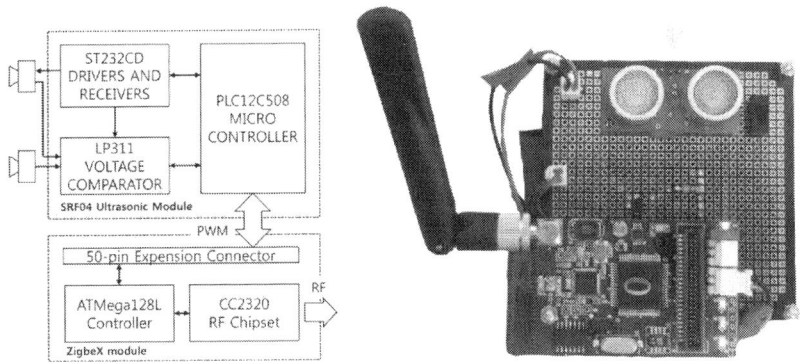

Figure 6. Vehicle detector hardware design.

The sensor module was developed with a Devantech SRF-04 ultrasonic sensor and its maximum detection range is up to 7 m, which is greater than that of other sensors. [7] The angle of the ultrasonic transmitter is 90 degree used on SRF series. Actually, the sensor received the reflected pulses in 40–45 degrees. An ultrasonic sensor with a range greater than that of the sensor used in this study could detect vehicles in additional lanes farther away from the sensor because the maximum detection range depends on the capabilities of the ultrasonic sensors made by different companies. Also sampling rate is determined to 20 Hz. The interval of ultrasonic bursts can be determined from minimum value 10 ms to maximum value 200 ms. Since the detection interval is determined to max value 200 ms, it can reduce a lot of power consumption of the sensor, but may not be able to detect a vehicle because speed of the vehicle is vary while the ultrasonic burst does not occur from sensor. Furthermore, the fast interval of bursts can detect the most of vehicle but increases the power consumption. Thus, we priority decide that the interval of ultrasonic bursts is 50 ms which is fast rate and reduces the power consumption. 50 ms is converted to 20 Hz sampling rate. The WUSM is manufactured without using battery power. However the ultimate goal of our system is the vehicle detecting system based on Wireless Sensor Network without restriction of the location. Thus, future work, we should research a new method such as sampling rate or sensing distance to consider the battery consumption of WUSM. The SRF-04 ultrasonic sensor in the proposed system consist of ST232CD drivers and receivers. The main module consists of a ZigBeX based ATMega128L controller and CC2320 radio transceiver [13].

EXPERIMENTS AND RESULTS

Experimental Environment

Figure 7 notes that the sensors are tested in the three place (red point) for experiments and the cases that the sensors are installed to various structure (the center of the figure). First of all, the system was installed and tested on a single one-lane road, a single two-lane downtown road, and a single two-lane highway in the city of Chuncheon in the province of Kangwon in the Republic of Korea.

Figure 7. Installed sensor on two-lane road, downtown (left) and highway (right).

The single two-lane downtown road has a speed limit of 70 km/h. It experiences heavy traffic flow and congestion in the daylight. The single one-lane downtown road has a speed limit of 60 km/h and is congested in the morning and evening. These sections were used to compare the accuracy of the system for the same traffic flow. The single two-lane highway has a speed limit of 80 km/h. This section was used to assess the accuracy of the proposed system in detecting vehicles passing through the sensor zone at high speed, in comparison to its accuracy in detecting vehicles on the two-lane downtown road. To detect vehicles on the road, ultrasonic sensors were installed on the side of the roads. There are various installation methods for our system. The structures installed on the side of the road (e.g., street trees, guardrails or streetlamp *etc.*) can be used as an installation place of WUSM as shown in the center of the Figure 7. Also, they can be fabricated and installed directly on the side of the road. The primary criterion considered in the performance evaluation was the accuracy of detection of the traffic flow and the location of vehicles in the multiple lanes. Tests were conducted during the daytime and nighttime.

Vehicle Detection Results

A WUSM installed on the side of a road can detect up to two lanes of traffic using one device. Therefore, the experiments had to be planned according to the number of lanes and the amount of traffic at various locations. To compare the accuracy of traffic volume determination in real environments, we measured the traffic volumes on individual one-lane and two-lane roads. In addition, tests were conducted on individual two-lane roads with different levels of traffic flow to assess the accuracy of the system in detecting and locating vehicles in the lanes.

Table 1 shows the data on the actual traffic flow and detected traffic flow on the single two-lane road. There is heavy traffic flow and congestion at rush hour and the error rates were 3.0% during daytime testing and 2.8% during nighttime testing. Note that there were 15 missing data items for the daytime testing, shown as missing data in the table. When there was heavy traffic on the road, if the vehicles in Lane 1 and Lane 2 passed through the range of the WUSM, which was installed on the side of the road, at the same time, the sensor could only detect the vehicle in Lane 2, closer to the sensor. In addition, if a vehicle changed lanes within the range of the WUSM, its location could not be detected, which resulted in missing data and over counting errors. Also if WUSM is obstructed by parked vehicles, it cannot detect any vehicles on the road. Therefore, any vehicles should not be parked within the range of the sensor.

Table 1. Detection result for single two-lane road downtown

Time	Actual Passing Vehicle (Vehicles)			Detected Vehicle (Vehicles)			Error Rate (Percent)			Type of Error (Vehicles)					
										Missing			Overcounting		
	Lane 1	Lane 2	Total	Lane 1	Lane 2	Total	Lane 1	Lane 2	Total	Lane 1	Lane 2	Total	Lane 1	Lane 2	Total
11:00–11:10	144	24	168	142	26	168	2.8	8.3	3.6	3	0	3	1	2	3
11:10–11:20	133	40	173	128	42	170	3.8	5.0	4.0	5	0	5	0	2	2
11:20–11:30	108	22	130	106	24	130	1.9	9.1	3.1	2	0	2	0	2	2
11:30–11:40	134	33	167	133	34	167	2.2	3.0	2.4	2	0	2	1	1	2
11:40–11:50	139	56	195	137	56	193	2.9	0.0	2.1	3	0	3	1	0	1
Total	658	175	833	646	182	828	2.7	5.1	3.0	15	0	15	3	7	10
20:00–20:10	114	37	151	115	37	152	2.6	0.0	2.0	1	0	1	2	0	2
20:10–20:20	121	41	162	124	41	165	4.1	0.0	3.1	1	0	1	4	0	4
20:20–20:30	112	31	143	114	33	147	1.8	12.9	4.2	0	1	1	2	3	5
20:30–20:40	148	32	180	148	30	178	0.0	12.5	2.2	0	3	3	0	1	1
20:40–20:50	157	21	178	157	22	179	2.5	4.8	2.8	2	0	2	2	1	3
Total	652	162	814	658	163	821	2.2	6.0	2.8	4	4	8	10	5	15

Table 2 show the data for the actual and detected traffic flow on the single one-lane road in the downtown areas. These results can be compared with the results shown in Table 1. This road is a single one-lane road, but it is as heavily trafficked and congested as the single two-lane road in the downtown area (Table 1) because many people use this road to go to the downtown area or the university. The data for the single one-lane road do not exhibit errors due to vehicles changing lanes or vehicles overlapping, as the data for the single two-lane road show. However, if vehicles stay in the sensor's range for long periods of time because of congestion, error due to noise data can result.

Table 2. Detection result for single lanes downtown.

| Time | Actual Passing Vehicle (Vehicles) | | | Detected Vehicle (Vehicles) | | | Error Rate (Percent) | | | Type of Error (Vehicles) | | | | | |
| | | | | | | | | | | Missing | | | Overcounting | | |
	Lane 1	Lane 2	Total	Lane 1	Lane 2	Total	Lane 1	Lane 2	Total	Lane 1	Lane 2	Total	Lane 1	Lane 2	Total
11:00–11:10	78	0	78	78	0	78	0	0	0	0	0	0	0	0	0
11:10–11:20	76	0	76	76	0	76	0	0	0	0	0	0	0	0	0
11:20–11:30	68	0	68	68	0	68	0	0	0	0	0	0	0	0	0
11:30–11:40	73	0	73	73	0	73	0	0	0	0	0	0	0	0	0
11:40–11:50	83	0	83	84	0	84	1.2	0	1.2	0	0	0	1	0	1
Total	378	0	378	379	0	379	0.3	0	0.3	0	0	0	1	0	1
20:00–20:10	77	0	77	78	0	78	1.3	0	1.3	0	0	0	1	0	1
20:10–20:20	86	0	86	86	0	86	0	0	0	0	0	0	0	0	0
20:20–20:30	80	0	80	80	0	80	0	0	0	0	0	0	0	0	0
20:30–20:40	91	0	91	92	0	92	1.1	0	1.1	0	0	0	1	0	1
20:40–20:50	85	0	85	86	0	86	1.2	0	1.2	0	0	0	1	0	1
Total	419	0	419	422	0	422	0.7	0.0	0.7	0	0	0	3	0	3

The detection results for the single two-lane highway with low traffic flow are shown in Table 3. This road is in the outer portion of the city and is lightly trafficked. The traffic volume at night is 50% lower than in the daytime. We know that the WUSM experienced frequent errors in monitoring the congested traffic flow on the single two-lane downtown road. It has been noted that the WUSM produced relatively high error, *i.e.*, 3% false rate, in detecting the vehicles in a congested flow. To address these types of problems and reduce the error rate, an additional WUSM could be installed on the other side of the road. The results for the single one-lane road downtown indicate relatively accurate detection of 0.3% and 0.7%, and the results for the two-lane highway indicate a low error rate because that highway has less traffic than the single two-lane downtown road.

Table 3. Detection result for single two-lane highway.

| Time | Actual Passing Vehicle (Vehicles) | | | Detected Vehicle (Vehicles) | | | Error Rate (Percent) | | | Type of Error (Vehicles) | | | | | |
| | | | | | | | | | | Missing | | | Overcounting | | |
	Lane 1	Lane 2	Total	Lane 1	Lane 2	Total	Lane 1	Lane 2	Total	Lane 1	Lane 2	Total	Lane 1	Lane 2	Total
11:00–11:10	38	56	94	35	56	91	7.89	0	3.1	3	0	3	0	0	0
11:10–11:20	61	54	115	61	55	116	0	1.8	0.8	0	0	0	0	1	1
11:20–11:30	71	46	117	68	48	116	4.2	4.3	4.2	3	0	3	0	2	2
11:30–11:40	53	44	97	53	47	100	0	6.8	3.0	0	0	0	0	3	3
11:40–11:50	44	38	82	42	38	80	4.5	0	2.4	2	0	2	0	0	0
Total	267	238	505	259	244	503	3.0	2.5	2.8	8	0	8	0	6	6
20:00–20:10	26	16	42	25	16	41	3.8	0	2.3	1	0	1	0	0	0
20:10–20:20	32	16	48	32	17	49	0	6.2	2.0	0	0	0	0	1	1
20:20–20:30	27	16	43	27	16	43	0	0	0	0	0	0	0	2	2
20:30–20:40	27	13	40	27	14	41	3.7	0	2.5	0	0	0	1	0	1
20:40–20:50	21	13	34	22	13	35	4.7	0	2.9	0	0	0	1	0	1
Total	133	74	207	133	76	209	2.3	1.4	1.9	1	0	1	2	3	5

Noise Filtering Results According to Window Mask Size

Noise data usually are generated as a vehicle approaches the sensor's angle or departs from it. When the vehicle departs from the sensor's angle, the sensor receives a small amount of noise because the surface area of the vehicle reflected by the carrier is also small. Therefore, the noise of various sizes can be generated according to characteristics of the vehicles in the measurement process and cannot be simply removed using a fixed filter window size. To find optimal filter size from the noise of various sizes, we experimented the error rates of the proposed system for filter window sizes of 3–9 cells. The experimental results can be used to determine the best filter size to most effectively eliminate noise by the vehicle detection system.

Figure 8 shows the error rate results according to the window size of the filtering mask. Each error rate is the average value of detection errors measured for each window size. The data measured in daytime and nighttime on three types of roads were used in the analysis. The results show that if a noise filter is not used, the error rate may be up to 16% However, if a noise filter is applied, the error rate decreases by 3%–6%, depending on the filter window size. The results show that the lowest value of 2.9% corresponds to a window size of 4 and that if the window size exceeds 4 the error rate increases.

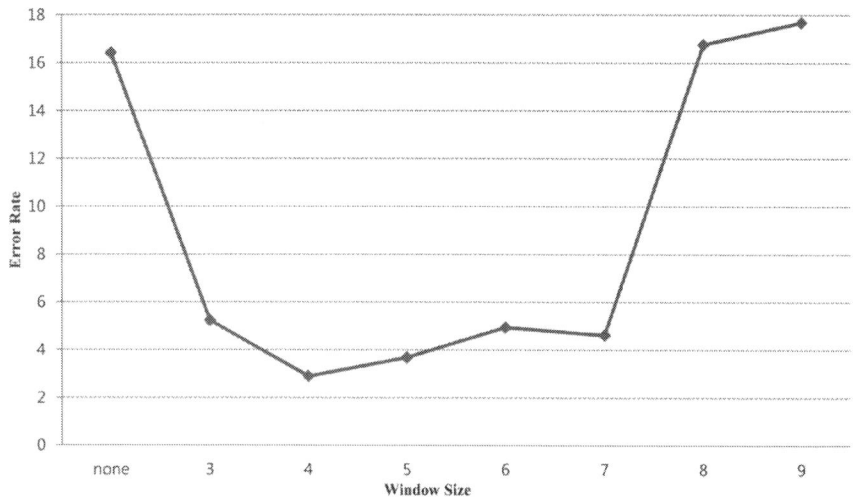

Figure 8. Overall error rate according to filter window size.

CONCLUSIONS

Traffic surveillance technologies can be classified as intrusive or non-intrusive technologies. Intrusive traffic sensors are installed within or across a pavement and have drawbacks that include high maintenance costs and disruption of traffic for installation, repair, or system reinstallation for road repairs or resurfacing. Non-intrusive sensors can be installed above or on the side of roads with minimum disruption to traffic flow and therefore offer the advantages of low installation costs and ease of maintenance. However, one of the disadvantages of non-intrusive sensors is that one must be installed for each lane.

In this paper, we propose a system for measuring vehicles using ultrasonic sensors mounted at the side of the road. The proposed algorithm consists of three parts can accurately detect vehicles in multiple lanes.

The proposed method was found to exhibit error rates of up to 3%, compared with actual traffic flow measurements, at three locations under two conditions (daytime and nighttime). However, there were very few differences in the results for a single two-lane downtown road and a rural highway road. Missing-vehicle errors occurred when multiple vehicles entered the sensor zone simultaneously. This happened more often on the heavily congested road. However, its effect was minimal in our experimental results.

The proposed ultrasonic sensor is small and can communicate with other sensors through a wireless network. Therefore, if sensors are installed on both sides of a road, vehicles can be detected more accurately than if a sensor is installed on only one side. In addition, because the WUSM is low in cost and easy to maintain, it can be used on country roads as well as complex metropolitan roads. Because of the low error rates achieved in vehicle detection and its low installation cost, the WUSM can contribute to vehicle detection as part of an ITS.

In future work, we plan to apply the sensor network system to a more extensive traffic environment. Therefore, the power management of our system will be considered so that the sensors can be maintained for a long time and remain able to transfer data.

This research was supported by Basic Science Research Program Through the National Research Foundation of Korea(NRF) funded by the Ministry of Education, Science and Technology (NRF-2013R1A1A2008811).

AUTHOR CONTRIBUTIONS

Soobin Jeon proposed the initial idea, designed the methodologies of lateral scanning-based ultrasonic sensing system, and wrote the text of the manuscript. All the studies of the manuscript were performed under the supervising of Eil Kwon and In bum Jung.

REFERENCES

1. Cheung, S.Y.; Varaiya, P.P. *Traffic Surveillance by Wireless Sensor Networks: Final Report*; California PATH Research Report UCB-ITS-PRR-2007-4. University of California: Berkeley, CA, USA, 2007. Available online: http://www.its.berkeley.edu/publications/UCB/2007/PRR/UCB-ITS-PRR-2007-4.pdf (accessed on 2 December 2014).

2. Festag, A.; Hessler, A.; Baldessari, R.; Le, L.; Zhang, W.; Westhoff, D. Vehicle-to-Vehicle and Road-Side sensor communication for enhanced road safety. Proceedings of the 15th World Congress on Intelligent Transport Systems, New York, NY, USA, 16–28 November 2008.

3. Jo, Y.T.; Kim, Y.; Jung, I.B. Variable Speed Limit to Improve Safety Near Traffic Congestion on Urban Freeways. *Int. J. Fuzzy Syst.* **2012**, *14*, 278–288. [Google Scholar]

4. Sun, Z.; Bebis, G.; Miller, R. On-road vehicle detection: A review. *IEEE Trans. Pattern Anal. Mach. Intell.* **2006**, *28*, 694–711. [Google Scholar]

5. Hussain, T.M.; Saadawi, T.N.; Ahmed, S.A. Overhead infrared sensor for monitoring vehicular traffic. *IEEE Trans. Veh. Technol.* **1993**, *42*, 477–483. [Google Scholar]

6. Fang, J.; Meng, H.; Zhang, H.; Wang, X. A low-cost vehicle detection and classification system based on unmodulated continuous-wave radar. Proceedings of IEEE Intelligent Transportation Systems Conference, Seattle, WA, USA, 30 September–3 October 2007; pp. 715–720.

7. Carullo, A.; Parvis, M. An ultrasonic sensor for distance measurement in automotive applications. *IEEE Sens. J.* **2001**, *1*, 143–147. [Google Scholar]

8. Kim, H.; Lee, J.H.; Kim, S.W.; Ko, J.I.; Cho, D. Ultrasonic vehicle detector for side-fire implementation and extensive results including harsh conditions. *IEEE Trans. Intell. Transp. Syst.* **2001**, *2*, 127–134. [Google Scholar]

9. Ma, F.; Zhang, Q.Y. The Ultrasonic Vehicle Flow Detection System Based on Ethernet. *J. Wuhan Univ. Technol.* **2007**, *49*, 182–184. [Google Scholar]

10. *Motor Industry Statistics*; Hyundai Motor Company: Seoul, Korea, 24 July 2008. Available online: http://hmc.vinyl.com/data/intro/report/2008_auto_industry.pdf(accessed on 2 December 2014)..

11. *Devantech SRF04 Ultra Sonic Ranger Module*; Devantech Ltd.: Norfolk, UK, 2006.

12. Ensminger, D.; Bond, L.J. *Ultrasonics: Fundamentals, Technologies, and Applications*; CRC Press: Boca Raton, FL, USA, 2011. [Google Scholar]

13. *ZigBee Specification*; ZigBee Alliance: San Ramon, CA, USA, 2006.

CITATION

Soobin Jeon, Eil Kwon and Inbum Jung, Traffic Measurement on Multiple Drive Lanes with Wireless Ultrasonic Sensors, doi:10.3390/s141222891

CHAPTER 6

An Analysis on Efficiency and Equity of Fixed-Time Ramp Metering

Ali Sercan Kesten[1,2], Murat Ergün[2], Tetsuo Yai[1]

[1]Department of Built Environment, Tokyo Institute of Technology, Tokyo, Japan
[2]Department of Civil Engineering, Istanbul Technical University, Istanbul, Turkey

ABSTRACT

Various traffic management strategies have been developed to alleviate the congestion on freeways. The equity issue has been considered as one of the major challenges for the implementation of some traffic control strategies, especially ramp metering. This paper presents a comparative evaluation of the efficiency and equity performance of a traffic control strategies namely Fixed Time Ramp Metering (FTRM). Instead of focusing on a single equity measure and/or indicator, different approaches to the equity concept are discussed and various equity measures are examined. The equity and efficiency performance of traffic control strategies are compared and evaluated by incorporating them into the simulated corridor. The Bosporus Bridge of Istanbul O-1 Freeway, Turkey is used as a test-bed for the simulation model and the control strategy is employed through microscopic traffic flow simulation software, VISSIM AG. The findings from the simulations show that the equity and efficiency properties of the network vary with the measures and indicators taken into account. The results also suggest that the trade-off between equity and efficiency can be observed for some measures, whereas regarding to other measures the trade-off is not validated.

INTRODUCTION

Freeways had been commonly recognized as to provide virtually unlimited mobility to road users, without any flow disturbance [1]. The constant increase of traffic demand, however, yields either recurrent congestion which occurs daily during rush hours or non-recurrent congestion which is defined as unexpected or unusual congestion [2]. The congested freeways within and around metropolitan areas degrades infrastructure utilization and causes increase in delays and Green House Gases (GHGs) emissions and decrease in average travel speed and safety [3-5]. The continuously increasing traffic congestion problem has led to application of various control strategies. Basically, these are formed by controlling the number of vehicles entering the freeway and/or by changing the speed limit of designated section along the freeway. Advanced urban traffic networks include both urban roads and freeways utilize control strategies like signal control, ramp metering, variable message signs and route guidance [6].

The efficiency of the traffic control is commonly analyzed through the measures of effectiveness such as; total travel time in hours, total delay time in minutes or hours, number of stops, average speed in km/h and total distance traveled in kilometers. The efficiency of traffic flow is essentially sensitive to demand levels, network topology, link geometry, and number of bottlenecks [7].

In practice mostly the links with higher traffic flows in the network are targeted for control where a maximum efficiency gain in travel time can imply a significant reduction of aggregate delays and generalized travel costs. On the contrary, this may trigger the increase of regional disparities [8].

In traffic control, in some cases it is observed that when efficiency is selected as the single most important measure, the impacts are not consistent with either environmental or social sustainable transportation principles [9]. Therefore, equity issues should be taken into account in traffic management and control. Before integrating equity issues to traffic control the broader definition of equity and specific meaning in traffic control should be put forward. Nevertheless, the definition and measurement of equity is one of the most controversial debates of humankind where a consensus is far over the horizon. There are some recent trials on integration of equity objectives in transportation planning and traffic engineering. The first studies of equity integration in transportation covered the fairness of transportation policies and later, road network design evaluated the equity issues [10]. The most common drawback of these researches is the single measure dependent nature. The authors of this paper believe that, various existing equity measures should be evaluated for the traffic control and the implications should be analyzed critically.

The goal of this paper is to present a brief review of efficiency and equity measures in transportation studies given the emphasis on traffic control and to analyze the efficiency and equity properties of FTRM strategy on a selected urban highway corridor.

The remainder of the paper is organized as fallows. The second section of this paper presents a brief review of efficiency and equity concepts in transportation planning giving a special emphasis on traffic control. FTRM methodology is introduced in Section 3. In Section 4, FTRM is applied in a traffic micro simulation environment a case study where the study site, data, calibration and simulation details are explained in detail. The equity and efficiency performance of FTRM are assessed in Section 5 and the conclusion section provides the implications of the efficiency and equity measures of FTRM control.

EFFICIENCY AND EQUITY CONCEPTS

Efficiency can be simply measured as a function of input and output of investments. In traffic control, the efficiency mostly considered as the gain of implementing a facility or a management strategy as the input where a reduction in total travel time, delay, number of stops, fuel consumption, emissions and accidents and/or increase in total traffic volume and speed as an output. In the case of ramp metering, speed management and route guidance, the investment is the cost of the installation of equipment, maintenance and operations. The return on the investment is the overall capacity increase and the indirect reduction in external costs. The significance and importance of measures can be varied in between and among the drivers, traffic authorities and the other peer groups of traffic. Therefore, the cost criterion may not be sufficient enough to draw a conclusion as it is the case in many of the studies in literature.

Equity is a complex idea that is not suitable to explain with simple formulations. It is mostly affected by cultural and societal values, by the specific types of goods i.e. divisible and non-divisible and conditional to given situation. Therefore, for a clear comprehension the contextual details are highly important. Equity has been one of the greatest topics debated as a matter of distribution of the prey. Thus, one of the simplest and shortest definitions attempts of equity could be justice and fairness in distribution of goods or rights ubiquitously. For the distribution among the group, firstly the group should be defined. The horizontal equity considers the group members have the same rights hence they deserve the same portion of distribution.

The concept of equity has been extensively examined in several disciplines e.g., health [11,12], political science [13], sociology [14], and economics [15,16]. Even though there is an increasing interest of utilizing equity concept in policy assessment, there is no solid and widely accepted framework available. There are numerous measures used in the literature, but a general consensus on the best measures to use for a particular case is still far away. Withal, there are also very good examples of equity analyses, one of them is the Peyton's work on conceptual evaluation of equity [17], and the other one is Cowell's systematic and delicate work on equity measurement [18]. Another significant review and framework is written by Marsh and Schilling which they analyzed equity measurement in facility location [19].

Approaches to Equity in Transportation

In the transportation literature, equity issues were generally concentrated on evaluation of the economic impacts of transportation policies. The picture has been changed especially with the studies concerning the distribution of impacts among various social groups in road pricing policies [20,21]. There are many equity analyses examples can be found in literature. Some of them are tabu lated in **Table 1**.

Equity Measures and Indicators

The equity of traffic control strategies are one of the biggest concern in practice. There are several indexes used in evaluation of equity issue in traffic control strategies mainly adapted from statistical or socio-economic models (welfare distribution). In this paper, the equity is rather used in the context of horizontal equity where in traffic control the drivers or alternatively cars should be treated equally[1].

Statistical measures examine the distribution of any variable in a given population. The most frequent use is the distribution investigation of income. Examples of these are; range, variance, measure of variation, log variance, Gini measure and Theil's entropy measure. Welfare measures are based on welfare economics and integrate equity concerns into a welfare function.

Table 1. Equity studies in transportation literature.

Equity Studies	Authors	Year
Mobility for Disadvantaged Groups	Stanley, et al.	2011
Women's Employment Access	Dobbs	2005
Public Funding Allocation	Chen	1996
Non-Drivers Accessibility	Case	2011
Inclusive Planning Analysis	Mann	2011
Transportation Improvement Benefit Distribution	Fruin and Sriraj	2005
Parking Requirement	Litman	2005
Transportation Cost Analysis	FHWA	1997
Transportation Cost Burdens	BLS	2000
Traffic Impacts	VTPI	2005
Planning Biases	Beimborn and Puentes	2003
Economic Opportunity	Gao and Johnston	2009
Transportation Pricing	TRB	2011
Spatial Analysis	Rodier, et al.	2010
Climate Change Emission Reduction	Lin	2008
VMT Reduction Strategies	Carlson and Howard	2010
Road Funding	Schweitzer and Taylor	2008
Fairness in a Car Dependent Society	SDC	2011
Right to Basic Transport	KOTI	2011

The axiomatic measures can be applied to the evaluation of inequality of any vector or distribution of observations, even to noneconomic data such as the distribution of the dispersion of pollutants or delay on a network. **Table 2** shows the measures are examined in this study and summarizes their properties.

In the measures given in Table; Y is a measure of welfare n is the number of observations on welfare $\overline{Y}$ is the mean level of welfare log Y is the mean level of log of welfare ε and α in Atkinson and Kolm measures respectively are the parameters that address inequality aversion.

TRAFFIC CONTROL STRATEGIES

Traditional control strategies use advanced technologies and more efficient procedures by integrating into the context of freeway management strategies that seek to manage, operate, and maintain expressways in an efficient and cost-effective manner [22]. The most effective control measures that are typically employed in freeway networks can be classified as ramp metering, speed management, route guidance and integrated control.

Table 2. A summary of equity indexes.

Measure	Definition	Properties[2]		
		T	SI	TI
Range	$R = Y_{max} - Y_{min}$			
Variance	$V = \dfrac{1}{n}\sum_{i=1}^{n}(Y_i - \bar{Y})^2$	+	−	−
Covariance	$c = \dfrac{\sqrt{V}}{\bar{Y}}$	±	+	−
Relative Mean Deviation	$M = \dfrac{1}{n}\sum_{i=1}^{n}\left|\dfrac{Y_i}{\bar{Y}} - 1\right|$	+	+	−
Logarithmic Variance	$\upsilon = \dfrac{1}{n}\sum_{i=1}^{n}\left(\log\left(\dfrac{Y_i}{\bar{Y}}\right)\right)^2$	−	+	−
Variance of Logarithms	$\upsilon_i = \dfrac{1}{n}\sum_{i=1}^{n}\left(\log\left(\dfrac{Y_i}{\bar{Y}}\right)\right)^2$	−	+	−
Gini	$G = \dfrac{1}{2n^2 \cdot \bar{Y}}\sum_{i=1}^{n}\sum_{j=1}^{n}|Y_i - Y_j|$	±	+	−
Theil	$T = \dfrac{1}{N}\sum_{i=1}^{N}\dfrac{Y_i}{\bar{Y}}\log\left(\dfrac{Y_i}{\bar{Y}}\right)$	+	+	−
Atkinson	$A_r = 1 - \left|\dfrac{1}{n}\sum_{i=1}^{n}\left[\dfrac{Y_i}{\bar{Y}}\right]^{1-r}\right|^{\frac{1}{1-r}}$	+	+	−
Kolm	$K_\alpha = \dfrac{1}{\alpha}\log\left(\dfrac{1}{N}\sum_{i=1}^{n}\exp\left(\alpha(\bar{Y} - Y_i)\right)\right)$	+	+	+

Ramp Metering

The use of traffic signals on-ramps to control the merging on freeways is called ramp metering. Ramp meters are installed to control the rate of vehicles moving into to the mainline traffic thus it prevents the critical volume of a freeway in order to control the demand and moreover, breaks the platoon of vehicles entering the freeway upstream of the signal to decrease the weaving phenomenon at the merge area. Ramp metering is projected to relieve or even eliminate congestion, ameliorate traffic flow conditions, safety and air quality, reduce total travel time and improve the performance measures, and regulate the demand in order to establish a stable freeway system [23].

Ramp metering is a well-known technique for freeways. In fact, various techniques of ramp control were used in the late 1950s and through 1970s in Japan and USA. By the early 1990s with the technological advancement both in computing and measurement techniques make more sophisticated ramp metering systems possible to analyze and implement. The specification of the metering which is the specific entrance allowance for vehicles from ramp to the freeway rate draws an important role in control success. An extensive literature reviews found on ramp metering algorithms and comparison of the performances of some of these algorithms are demonstrated in [24-27].

Three different metering operations can be defined according to the control logic as: fixed-time, local traffic responsive and coordinated traffic responsive. A fixed time ramp-metering control uses historical traffic data and a time-of-day basis [28]. Local traffic responsive ramp-metering strategies use the measurements of traffic flow and the metering rate is based on prevailing traffic conditions in the vicinity of the ramp. The most prominent examples of the local ramp-metering strategies are the demand capacity (DC), the occupancy (OCC) strategies and (ALINEA) strategy [29]. Local traffic responsive metering algorithms regardless of type of controller which can be either linear [30], artificial neural network [31] or Fuzzy-logic [32] are performed well without considering the system wide optimization. The coordinated traffic responsive ramp metering aims optimization of the performance of entire freeway facility. Fixed time and/or local traffic responsive control approaches could be used in concert with the coordinated traffic responsive control approach by predicting the traffic conditions. Coordinated ramp-metering strategies benefits the measurements from the entire network to control all metered ramps. Some studies [33,34] stated that coordinated traffic responsive strategies are more efficient when the demand is extremely high. Contrary in some studies [35,36] coordinated control algorithms are obtained not superior to the local ramp metering strategies. The main drawback of coordinated traffic responsive ramp metering approach is the complex and costly nature to realize.

Fixed-Time Ramp Metering

Historically, the solution of the conflicts between the multidirectional flows of traffic is sought by considering the allocation of saturation, time or delay among all movements. The use of traffic signal establishes an orderly movement of traffic and increases the capacity and safety of intersections thoroughly. The design process of timing plans for signalized intersections in Highway Capacity Manual [37] treats the traffic merely as static volumes of conflicting movements that require right-ofway alternatively. With a given phase sequence and phase groups, the method can determine how much green time within a cycle will be allocated to each phase, or the green splits. One fundamental difference of these methods is the design logic to allocate green splits; and these logic will affect how efficient and equitable a timing plan can be. Three major logics have been developed are; equal-saturation strategy [38] where the green time is determined in such a way that the phase duration will be proportional to its critical Volume/Capacity (V/C) ratio, delay minimization strategy [39] which is a policy that minimizes the total intersection delay and the capacity maximization policy [40] maximizing the intersection capacity through balancing the traffic pressures of conflicting approaches where the pressure is defined as the product of the approach capacity and its average delay for each approach link.

Likewise traffic lights, ramp metering control utilize traffic signals at freeway on ramps or freeway interchanges to manage the rate of vehicles entering the freeway. However, there is no established method for the specification of optimum cycle time for fixed time ramp metering. Therefore, the traditional analytical models developed for fixed time intersection control are examined and capacity maximization approach is modified for fixed time ramp metering simulation experiment.

Ramp metering algorithms aim to set the allowable ramp flow value r (in veh/h) which can be basically defined as (cr/3600) vehicles where c denotes cycle time in seconds [41]. Traffic lights are operated on the basis of a traffic cycle consisting of a green phase T_G, an amber phase T_A, a red phase T_R, and a red-amber phase T_{AR} which are adjusted in seconds such that:

$$c = T_G + T_A + T_R + T_{AR} \tag{1}$$

In this study, the number of vehicles that the signals allow off the ramp is calculated as the difference between the actual demand at the bottleneck, more specifically the sum of the mainline and ramp flows) and the pre-specified capacity of the road. The most critical point is the specification of the bottleneck capacity since it varies over time. Nevertheless, in most cases bottlenecks are also considered as having the same capacity as basic freeway segments which takes values between 1800 and 2200 veh./h/lane. The excess demand (D_e) would be determined from:

$$D_e = D_a - C \tag{2}$$

where D_a is the actual demand (in veh./h) including ramp and mainline flows and C (in veh./h) is the capacity of the downstream section of the bottleneck. Resulting from (4), the admissible ramp flow value (r) would be:

$$r = C - D_e \tag{3}$$

The translation of the ramp flow value r into a corresponding green phase under a full traffic cycle plan, where the traffic cycle c is always equals to the metering period, would be based on the green ratio (f):

$$f = r/C \tag{4}$$

Therefore, the green time leads to:

$$T_G = cf \tag{5}$$

TRAFFIC MICRO-SIMULATION MODELING

The Bosporus Bridge in Istanbul, Turkey, is the first of the two highway crossings connecting Asia and Europe over the Bosporus Strait. In this study the traffic from Asia to Europe direction along O1 route, schematically shown in **Figure 1**, is selected as the study site.

Study Site and Data

The corridor investigated has approximately 7 km of length, where there are 6 entrance ramps and 2 exit ramps up to the Bosporus Bridge. There are 4 main junctions entering/exiting to/from O1 Route and the bottlenecks are mostly occurring at around the downstream sections of the junctions (S4, S7, S11 and S13) due to the merging flow. In morning hours, especially weekdays the queue length may reach kilometers long and the average speed on the corridor can decrease down to 5 km/h which indicates a complete hyper-congestion.

The data is collected at the Istanbul Traffic Control Center at 14th of March, 2011. The traffic flow is observed from 6:00 a.m. to 11:30 a.m. through video recordings. Later, manual counts aggregated to 15 min. and inserted to commercial spreadsheet program (Microsoft Excel). The volumes at ramps are high between 6:00 a.m. and 7:00 a.m. especially most of the vehicles are medium type vehicles (minibus or midibus) which are used as service vehicles. Service vehicles could be classified as a special type of car sharing model which is mostly provided by companies free of charge to their employees. The hourly volumes of some ramps (S9, S10 and S12) even have higher volumes per lane than the mainline for a short time period.

The on-ramps have two different geometries as; 5 m width single lane and dual lanes with 3.5 m width. However, a virtual lane occurs at every single ramp during congested hours. The congestion starts at 6:45 m. and the flow decreases down to 400 veh./h/lane at the bottleneck downstream sections. It is observed that, once the breakdown occurs along the O1 Route, the congested flow remains invariant regardless of the time of the day which is verified by Sahin et al. [42].

The speeds are also calculated through the recorded videos despite the measurement is not based on an approved method. The average speed is represented by randomly taken cars (medium, heavy) for a time period.

The speed profiles are only used in visual conformity check and not considered for calibration purposes due to possible measurement errors.

However, the results indicate that the speed profiles at ramps are relatively lower than mainline speed. The average mainline speed decreases down to 20 km/h after 6:45 a.m. and oscillates between 30 to 40 km/h for automobiles afterwards. The average ramp speed is around 20 km/h for automobiles and after 9:00 a.m. the speed increases to 30 km/h.

Calibration and Simulation

There are two ways to evaluate the performance of ramp metering systems: field operational tests and computer simulations. Although field tests provide more realistic results, due to the high costs and time consuming nature, traffic simulation studies are becoming more popular. In this study, widely accepted, discrete, stochastic, time step based microscopic traffic flow simulation software, VISSIM, is employed to test the performance of control strategies and compare their performances.

VISSIM utilizes psychophysical car following models which combines a perceptual driver behavior model with a vehicle dynamic model (1974, 1999) [43-45].

The study corridor is simulated for the morning peak hours which start from 6:30 a.m. to 9:30 a.m. and performance measurement interval is selected as 15 minutes.

The traffic composition and the priorities at the ramp weaving areas are set through the analyses of video recordings. It is observed that the vehicles entering to the mainline are more aggressive than the vehicles cruising on the right most lanes. Therefore the priority is given to the ramp flows over mainline flows in simulation. It is also watched that, if there is an enough gap the drivers tend to change the left most lane within the minimum possible distance. Typically, the drivers are highly aggressive and breaking and acceleration values are taken higher than the default values. Lane changing is also highly strong in Istanbul traffic and drivers are frequently cutting in and overtaking. The car following model is selected as Wiedmann 1999, which has ten driver behavior parameters labeled CC0-CC9. Several driver behavior parameters are reported to have significant impacts on roadway capacity and speed profiles thus, the parameters need to be optimized to attain the visual conformity and numerical correlation between the observation and simulation [46].

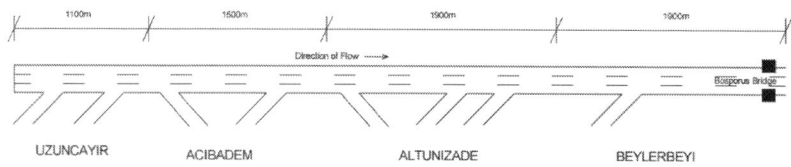

Figure 1. Segments of study corridor.

In the model calibration process, model parameters are altered until a qualitative and a quantitative balance between the simulation and the observation is reached. Traditionally, calibration requires several runs based on engineering judgment and experience. A three step calibration procedure is applied in this study, which are; calibration of driving behavior models, OD estimation and model fine-tuning.

The mean target headway and driver reaction time, which are the key user specified parameters in the car-following and lane changing models, can drastically influence overall driver behaviors of the simulation [46]. The calibrated values of the two parameters are 0.6 sec and 1.5 sec in this study. The calibration of ten parameters in car following model could be performed through some optimization techniques in order to achieve the most representative model. However, this is not the focus of this paper. Likewise, the local arterial roads are not included in the studied network hence, route choice is not considered in this calibration process.

In this study, the observations of Chu et al. [47] confirmed that the precise geometry of merging angle and connector link length have an impact of simulation accuracy. Proportion of each vehicle type, vehicle characteristics and performance, such as the acceleration and deceleration rate, driving restrictions, the speed limits and the driving lane restriction, the look back distance at merging and bifurcation weaving area, the priorities and traffic flow bases on conflicting areas also effects the simulation results.

In calibration process GEH index [48] is often used to test the relative difference between observed (Q_o) and simulated (Q_s) link volumes. GEH formula can be calculated with equation (6).

$$GEH = \sqrt{\left(2(Q_o - Q_s)^2\right)\big/(Q_o + Q_s)}$$

(6)

The simulation model is acceptable for the GEH scores are smaller than 5 in 85% of the links and smaller than 4 for the sum of all link counts. The GEH values are below the limit values thus the simulation model developed is considered as representative.

In order to determine the optimal cycle time and green time for fixed time ramp metering control and examine the cycle time duration effects on network performance, a set of simulation experiments is designed. At each ramp, the green times are calculated for the average flows of entire simulation period by varying the signal cycle time from 5 to 20 seconds.

EFFICIENCY AND EQUITY PERFORMANCE OF TRAFFIC CONTROL STRATEGIES

The objective of the freeway traffic control process is to optimize a performance index that mostly consists of efficiency measures. Performance index can be stated to minimize the travel times, delays, number of stops, or some other parameters such as fuel consumption and environmental pollution or in a more social context the optimization temporal and spatial of equity along the network or a more comprehensive objective that considers all the aspects with suitable weighting. However, only the efficiency properties are investigated for each control strategies in this study.

Efficiency Performance

The first performance measure is selected as total travel time. The Total travel time is calculated in hours for all active and arrived vehicles. In addition to the total travel time, the total delay in hours, the total number of stops and the average speed in km/h are evaluated by averaging values of 15 min. intervals for each simulation run.

Table 3 compares the performance measure of control strategies investigated. It shows that all the traffic control strategies significantly increase the network performance.

According to the results obtained, the best network performance achieved for FTRM control at 15 sec cycle time.

When the 15 sec cycle time control is compared with no control case, it can be seen that the total travel time, the total delay and the number of stops decreased by 32%, 60% and 80% respectively and the actual average speed increase from 29.2 km/h to 44.7 km/h.

Equity Performance

The results of the equity performances of on-ramps, mainline and the corridor is given for the indicators of spot speed, space mean speed and delay in **Table 4**.

From the evaluation results of on-ramps, mainline and corridor wide performances, it is obvious that the FTRM increased the inequality on ramps when the spot speeds are taken into account.

Table 3. Efficiency performance of FTRM.

Measures of Efficiency	No Control	FTRM
Total travel time [h]	4942	3368
Total delay time [h]	2910	1190
Number of stops	411,772	81,634
Average speed [km/h]	29.2	44.7
Total Distance Traveled [km]	144,406	150,460
Number of vehicles in the network	2189	1065
Number of vehicles that have left the network	26,696	27,718
Total stopped delay [h]	374	52
Average delay time per vehicle [s]	363	149
Average stopped delay per vehicle [s]	47	7
Average number of stops per vehicles	14	3

Table 4. Equity performance of FTRM control.

Measure	NO CONTROL (Spot Speed, km/hr)			FTRM (Spot Speed, km/hr)			FTRM (Space Mean Speed, km/hr)			FTRM (Delay, min)		
	ON-RAMPS	MAIN LINE	CORRI-DOR	ON-RAMPS	MAIN LINE	COR-RIDOR	ON-RAMPS	MAIN LINE	CORRI-DOR	ON-RAMPS	MAINLI NE	CORRI-DOR
Range	27.55	14.25	16.86	41.65	22.78	27.63	8.37	6.98	7.15	3.1	6.4	3.19
Variance	102.1	24.2	37.3	191.6	66.4	89.3	4.2	3.7	3.4	0.7	3.7	0.9
Coefficient of Variance	0.29	0.11	0.14	0.32	0.16	0.19	0.05	0.04	0.04	0.02	0.03	0.02
Relative Mean Deviation	0.26	0.09	0.12	0.26	0.14	0.16	0.04	0.03	0.03	0.01	0.03	0.01
Logarithmic Variance	1.3E-02	2E-03	3.E-03	2.2E-02	5E-03	7E-03	4.8E-04	2.7E-04	2.7E-04	4.9E-05	1.8E-04	4.9E-05
Variance of Logarithms	1.805	2.028	2.051	1.985	2.166	2.149	1.98	2.177	2.133	2.183	2.373	2.306
GINI	0.143	0.06	0.069	0.188	0.09	0.11	0.024	0.02	0.019	0.009	0.016	0.008

The aversion parameter of ε is taken as 5 in Atkinson Index and α in Kolm Index calculation is taken as 0.025.

However, when equity of the on-ramps is compared through space mean speed and delay as indicators, the results are dramatically favoring the equity of no-control policy. The possible explanation for this controversy is the measurement methodology of the indicators. The equity measures are basically calculating the distribution of differences from the mean value. The spot speeds are relatively stable in no control case, which actually show the heavy congestion on ramps. With the help of FTRM, the on-ramps have a tendency to fluctuate because of no queue control mechanism which yields sudden spot speed changes on ramps. Therefore, the selection of spot speed detection location gains an extreme importance due to this fluctuating nature of FTRM control. On the other hand, the space mean speed and delay indicators create even more equal tableau than no-control case where the indicator calculations are based on spatial measurements.

CONCLUSIONS

This study demonstrates the potential of implementing traffic control strategies in alleviating the traffic congestion on an urban freeway. The fixed time ramp metering control is analyzed as a traffic control strategy. The traffic simulation network is modeled in a traffic micro simulation environment and the traffic model is calibrated for the analyses.

The FTRM is successfully increased the effectiveness of the traffic flow referring the total travel time, the total delay, the number of stop and average speed. The prevailed results lead that; there is an optimum cycle time that can be determined for each on ramp considering the bottleneck downstream capacity and flows. The short cycle lengths have a tendency to increase the start-up lost times and limits the merging rate, therefore the delay increases rapidly. Long cycle lengths allows platoon of vehicles entering the mainline which also contributes an increase in delay. The model shows that there is an optimum cycle length obtaining the best performance values for the merging section. The result of study is highly congruent with previous findings on sensitivity of delay to cycle length for intersections exhibited on 16-16 at Highway Capacity Manual [37].

Another important output of this study is the results accord with the existing literature suggesting that the ramp control brings equity concerns for ramp users when the spot speeds are taken into account [49]. From the results, it is also observed that the equity of ramps is worsened with the control strategy when it is compared to the equity of mainline and overall corridor.

Finally, the change in magnitude of equity with respect to the measure and indicator taken into account could be vast. Thus, first the equity indicators and measures should be evaluated carefully and then the performance should be interpreted accordingly to make a conclusive judgment.

As evaluation results indicated the controversial nature of selection and evaluation of efficiency and equity, the future research would be continuing on the examination of other traffic control strategies such as speed management, integrated control as well as dynamic ramp metering control.

ACKNOWLEDGEMENTS

The authors would like to thank PTV Japan for providing the license key of VISSIM and their continual support during the research. Ali Sercan Kesten acknowledges the support of ITU and TiTech, and his PhD research is supported by Ministry of Education, Culture, Sports, Science & Technology of Japan.

REFERENCES

1. M. Papageorgiou, C. Diakaki, V. Dinopoulou, A. Kotsialos and Y. Wang, "Review of Road Traffic Control Strategies," IEEE Proceedings, Vol. 91, No. 12, 2003, pp. 2043-2067.

2. M. E. Hallenbeck, J. M. Ishimaru and J. Nee, "Measurement of Recurring versus Non Recurring Congestion," Washington State Transportation Center, Olympia, 2003.

3. J. Kwon, M. Mauch and P. Varaiya, "The Components of Congestion: Delay from Incidents, Special Events, Lane Closures, Weather, Potential Ramp Metering Gain, and Excess Demand," Proceedings of 85th Annual Meeting of the Transportation Research Board (CD-ROM), Washington DC, 2006.

4. M. Barth and K. Boriboonsomsin, "Real-World CO_2 Impacts of Traffic Congestion," Paper for the 87th Annual Meeting of Transportation Research Board, Washington DC, 2008, p. 10

5. T. F. Golob, W. W. Recker and V. M. Alvarez, "Freeway Safety as a Function of Traffic Flow," Accident Analysis and Prevention, Vol. 36, No. 2004, pp. 933-946.doi:10.1016/j.aap.2003.09.006

6. G. Pesti, P. Wiles, R. L. Cheu, P. Songchitruksa, J. Shelton and S. Cooner, "Traffic Control Strategies for Congested Freeways and Work Zones," Final Report FHWA/ TX-08/0-5326-2, Texas Transportation Institute, College Station, 2007.

7. J. C. Williams, "Macroscopic Flow Models," In: N. Gartner, C. J. Messer and A. K. Rathi, Eds., Traffic Flow Theory, Oak Ridge National Laboratory, Oak Ridge, Chapter 6, 1997, pp. 1-31.

8. J.-F. Thisse, "How Transport Costs Shape the Spatial Pattern of Economic Activity," Working Paper, OECD/ ITF, London, 2009.

9. D. J. Rosa, "Sustainability and Infrastructure Resource Allocation," Journal of Business & Economics Research (JBER), Vol. 7, No. 9, 2011, pp. 71-76.

10. B. Santos, A. Antunes and E. J. Miller, "Integrating Equity Objectives in a Road Network Design Model," Transportation Research Record: Journal of the Transportation Research Board, Vol. 2089, No. 1, 2008, pp. 35-42. doi:10.3141/2089-05

11. P. Konings, S. Harper, J. Lynch, A. R. Hosseinpoor, D. Berkvens, V. Lorant and N. Speybroeck, "Analysis of Socioeconomic Health Inequalities Using the Concentration Index," International Journal of Public Health, Vol. 55, No. 1, 2010, pp. 71-74.doi:10.1007/s00038-009-0078-y

12. A. Pinto, H. Manson, B. Pauly, J. Thanos, A. Parks and A. Cox, "Equity in Public Health Standards: A Qualitative Document Analysis of Policies from Two Canadian Provinces," International Journal for Equity in Health, Vol. 11, No. 1, 2012, p. 28. doi:10.1186/1475-9276-11-28

13. J. L. Hall "The Distribution of Federal Economic Development Grant Funds: A Consideration of Need and the Urban/Rural Divide," Economic Development Quarterly, 2010. doi:10.1177/0891242410366562

14. J. L. Hall, "The Distribution of Federal Economic Development Grant Funds: A Consideration of Need and the Urban/Rural Divide," Economic Development Quarterly, 2010. doi:10.1177/0891242410366562

15. J. P. Chavas, "Equity Considerations in Economic and Policy Analysis," American Journal of Agricultural Economics, Vol. 76, No. 5, 1994, pp. 1022-1033.

16. B. Hansen, "The Political Economy of Poverty, Equity, and Growth: Egypt and Turkey," Oxford University Press, Oxford, 1991. doi:10.2307/1243386

17. H. P. Young, "Equity: In Theory and Practice," Princeton University Press, Princeton, 1995.

18. F. Cowell, "Measuring Inequality," Oxford University Press, Oxford, 2011.

19. M. T. Marsh and D. A. Schilling, "Equity Measurement in Facility Location Analysis: A Review and Framework," European Journal of Operational Research, Vol. 74, No. 1, 1994, pp. 1-17. doi:10.1016/0377-2217(94)90200-3

20. J. M. Viegas, "Making Urban Road Pricing Acceptable and Effective: Searching for Quality and Equity in Urban Mobility," Transport Policy, Vol. 8, No. 4, 2001, pp. 289- 294.doi:10.1016/S0967-070X(01)00024-5

21. A. Karlström and J. P. Franklin, "Behavioral Adjustments and Equity Effects of Congestion Pricing: Analysis of Morning Commutes during the Stockholm Trial," Transportation Research Part A: Policy and Practice, Vol. 43, No. 3, 2009, pp. 283-296.doi:10.1016/j.tra.2008.09.008

22. L. Jacobson, J. Stribiak, L. Nelson and D. Sallman, "Ramp Management and Control Handbook," Federal Highway Administration, Washington DC, 2006.

23. H. M. Zhang, T. Kim, X. Nie, W. Jin, L. Chu and W. Recker, "Evaluation of On-Ramp Control Algorithms," California PATH Research Report, University of California, Berkeley, 2001.

24. M. Hasan, "Evaluation of Ramp Control Algorithms Using a Microscopic Simulation Laboratory," Master Thesis, Massachusetts Institute of Technology, Cambridge, 1999.

25. J. R. Scariza, "Evaluation of Coordinated and Local Ramp Metering Algorithms Using Microscopic Traffic Simulation," Master Thesis, Massachusetts Institute of Technology, Cambridge, 2003.

26. R. Horowitz, X. Sun, L. Munoz, A. Skabardonis, P. Varaiya, M. Zhang and J. Ma, "Design, Field Implementation and Evaluation of Adaptive Ramp Metering Algorithm: Final Report, California PATH Research Report," UCB-ITS-PRR-2006-21, 2006.

27. K. Bogenberger and A. D. May, "Advanced Coordinated Traffic Responsive Ramp Metering Strategies," California PATH Working Paper UCB-ITS-PWP-99-19, 1999.

28. M. Papageorgiou, and A. Kotsialos, "Freeway Ramp Metering: An Overview," IEEE Transactions on Intelligent Transportation Systems, Vol. 3, No. 4, 2002, pp. 271- 281.

29. M. Papageorgiou, H. Haj-Salem and F. Middelham, "ALINEA Local Ramp Metering: Summary of Field Results," Transportation Research Record, No. 1603, 1998, pp. 90-98.

30. M. Papageorgiou, H. Haj-Salem and J. M. Blosseville, "ALINEA: A Local Feedback Control Law for On-Ramp Metering," Transportation Research Record, No. 1320, 1991, pp. 58-64.

31. H. M. Zhang and S. Ritchie, "Freeway Ramp Metering Using Artificial Neural Networks," Transportation Research Part C: Emerging Technologies, Vol. 5, No. 5, 1997, pp. 273-286.doi:10.1016/S0968-090X(97)00019-3

32. C. Taylor, D. Meldrum and L. Jacobson, "Fuzzy Ramp Metering—Design Overview and Simulation Results," Transportation Research Record, Vol. 1634, 1998, pp. 10-18.doi:10.3141/1634-02

33. C. Taylor and D. Meldrum, "Evaluation of a Fuzzy Logic Ramp Metering Algorithm: A Comparative Study among Three Ramp Metering Algorithms Used in the Greater Seattle Area WSDOT Technical Report WA-RD 481.2," 2000.

34. S. Ahn, R. L. Bertini, B. Auffray, J. H. Ross and O. Eshel, "Evaluating the Benefits of a System-Wide Adaptive Ramp-Metering Strategy in Portland," Journal of Transportation Research Board, Transportation Research Record, Vol. 2007, No. 1, 2012, pp. 47-56.

35. L. Chu, H. X. Liu, W. Recker and H. M. Zhang, "Performance Evaluation of Adaptive Ramp-Metering Algorithms Using Microscopic Traffic Simulation Model," Journal of Transportation Engineering, Vol. 130, No. 3, 2004, pp. 330-338.doi:10.1061/(ASCE)0733-947X(2004)130:3(330)

36. K. Ozbay, I. Yasar and P. Kachroo, "Modeling and PARAMICS Based Evaluation of New Local Freeway Ramp Metering Strategy That Takes into Account Ramp Queues," Transportation Research Record, No. 1867, 2004, pp. 89-97.

37. "Highway Capacity Manual," Transportation Research Board, Washington DC, 2000.

38. F. V. Webster, "Traffic Signal Settings," Road Research Technique Paper No. 39, Road Research Laboratory, London, 1958.

39. R. B. Allsop, "SIGSET: A Computer Program for Calculating Traffic Capacity of Signal-Controlled Road Junctions," Traffic, Engineering and Control, Vol. 12, No. 2, 1971, pp. 58-60.

40. M. Papageorgiou and I. Papamichail, "Handbook of Ramp Metering. Deliverable 7.5," Report for the European IST Office, Brussels, 2007.

41. M. J. Smith and T. Van Vuren, "Traffic Equilibrium with Responsive Traffic Control," Transportation Science, Vol. 27, No. 2, 1993, pp. 118-132. doi:10.1287/trsc.27.2.118

42. I. Sahin and G. Akyildiz, "Bosporus Bridge Toll Plaza in Istanbul, Turkey: Upstream and Downstream Traffic Features," Transportation Research Record: Journal of the Transportation Research Board, Vol. 1910, No. 1, 2005, pp. 99-107.

43. R. Wiedemann, "Simulation des Straßenverkehrsflusses," Schtifienreihe des Instituts fur Verkehrswesen der Universität Karlsruhe, Heft 8, 1974 (in German).

44. R. Wiedemann, "Modeling of RTI-Elements on Multilane Roads," Advanced Telematics in Road Transport Edited by the Commission of the European Community, Vol. DG XIII, 1991.

45. PTV, "VISSIM Version 5.40 User Manual," PTV Planug Transport Verkehr AG, Stumpfstraße 1 D-76131 Karlsruhe, Germany, 2011.

46. N. Lownes and R. Machemehl, "Vissim: A Multiparameter Sensitivity Analysis," Winter Simulation Conference, Monterey, 3-6 December 2006, pp. 1406-1413.

47. L. Chu and X. Yang, "Optimization of the ALINEA Ramp-Metering Control Using Genetic Algorithm with Micro-Simulation," Transportation Research Board 82nd Annual Meeting, Washington DC, 12-16 January 2003, pp. 22-24.

48. K. Chu, L. Yang, R. Saigal and K. Saitou, "Validation of Stochastic Traffic Flow Model with Microscopic Traffic Simulation," IEEE International Conference on Automantion Science and Engineering, Trieste, 24-27 August 2011, pp. 672-677.

49. L. Zhang and D. Levinson, "Ramp Metering and Freeway Bottleneck Capacity," Transportation Research Part A, Vol. 44, No. 4, 2010, pp. 218-235.doi:10.1016/j.tra.2010.01.004.

CITATION

A. Kesten, M. Ergün and T. Yai, "An Analysis on Efficiency and Equity of Fixed-Time Ramp Metering," *Journal of Transportation Technologies*, Vol. 3 No. 2A, 2013, pp. 48-56. doi: 10.4236/jtts.2013.32A006.

CHAPTER 7

An Approach to an Intersection Traffic Delay Study Based on Shift-Share Analysis

Jianfeng Xi [1], Wei Li [1], Shengli Wang [1] and Chuanjiu Wang [2]

[1]College of Traffic, Jilin University, Changchun 130022, China;

E-Mails: liwei1337@gmail.com(W.L.); wsl20101130@163.com (S.W.)1

[2]Chongqing CMTraffic Tech. Co., Ltd., Chongqing 400060, China;

E-Mail: chuanjiu@126.com

ABSTRACT

Intersection traffic delay research has traditionally placed greater emphasis on the study of through and left-turning vehicles than right-turning ones, which often renders existing methods or models inapplicable to intersections with heavy pedestrian and non-motorized traffic. In the meantime, there is also a need for understanding the relations between different types of delay and how they each contribute to the total delay of the entire intersection. In order to address these issues, this paper first examines models that focus on through and left-turn traffic delays, taking into account the presence of heavy mixed traffic flows that are prevalent in developing countries, then establishes a model for calculating right-turn traffic delay and, last, proposes an approach to analyzing how much each of the three types of traffic delay contributes to the total delay of the intersection, based on the application of shift-share analysis (SSA), which has been applied extensively in the field of economics.

INTRODUCTION

Intersection delay studies play an important role in traffic planning, signal control design and determining the level of service (LOS) at signalized intersections. The study of intersection delay has traditionally focused more on through and left-turn traffic, which may have its merit when the intersection of concern has very little pedestrian traffic. However, in cases like China or any other developing country with a dense urban population, where there are constant flows of pedestrian and non-motorized traffic conflicting with right-turning vehicles, existing methods or models are frequently inadequate in assisting delay analysis of urban intersections.

One major reason for the lack of research on right-turn delay is due to the fact that most signalized intersections do not warrant control on right-turning traffic, which, again, presents a special problem for developing countries, like China and India, where there are a large number of "first-generation" new drivers who lack either the awareness of or the compliance to basic rules of the road, often both. On the other hand, there are also constant violations of traffic rules from the pedestrian side. These two combined often lead to increased total delay at urban intersections and, as a consequence, lowering of the level of service (LOS), as well.

Another area of traffic delay study is the investigation into the relations between different types of delay at an intersection and how they each contribute to the total delay of the intersection. To propose an approach to addressing the problem, this papers introduces shift-share analysis (SSA), which is used in regional science, political economy and urban studies, to determine what portions of regional economic growth or decline can be attributed to national, economic industry and regional factors. SSA is used here to analyze how much each of the three types of delay (i.e., left-turn, through and right-turn delay) contributes to the total delay of the intersection. The introduction of SSA into the study of traffic delay helps broaden the options for determining the causes and understanding the patterns of intersection delay and also provides a new tool for intersection traffic analysis and design.

LITERATURE REVIEW

Traffic delay is one of the major criteria in determining the level of service or the effectiveness of traffic operations at signalized intersections [1,2]. The overall level of congestion, driver comfort, vehicle gas consumption and the average loss of travel time, among others, can all be attributed to traffic delay. Traffic delay is also an area of concern in traffic planning, signal design and traffic control management [3,4]. Since the 1950s, intersection delay analysis has been one of the focuses of study for transportation professionals [5,6,7]. Many delay models and evaluation methods were introduced [8,9], a significant number of which were addressing the modeling of through and left-turning vehicular traffic at signalized intersections. The results have been mostly positive in reflecting real-world traffic operations [10].

To better investigate and understand delays involving right-turning vehicles, some researchers took the traditional approach of establishing right-turn delay models [11,12]. Zhang et al. suggested that right-turn delay was caused mostly by pedestrian interference, but did not present situations in which pedestrians are compliant with traffic controls [13]. Qureshi et al. introduced a method of calculating the average delay for right-turning vehicles on red through the use of queuing accumulation polygons (QAP), which was developed from queuing theory [14]; however, situations in which vehicles come under interference from pedestrians were not discussed. Guo et al. put forward a model for calculating delay caused by pedestrians passing through right-turning traffic and made an improvement to delay analysis by bringing vehicle types into consideration [15]. Zhou et al. introduced a choice model based on individual pedestrian crossing behavior [16]. Tao et al. studied vehicle headway and established a mixed distribution model [17]. Zhao et al. conducted a regression-based pedestrian adjustment factor analysis that involves right-turn traffic [18].

Most of the above-mentioned models analyze each type of delay separately from the influences from the others and lack efforts in examining the relations among different types of delay. Moreover, total delay alone does not reflect the delay changes in different ways (legs) and lanes of the intersection. Therefore, it is hard for such analyses to draw deeper conclusions on the true causes and patterns of intersection delay.

A shift-share analysis attempts to identify the sources of regional economic changes [19,20], which helps identify industries in which a regional economy has competitive advantages over the larger economy [21,22]. SSA can be used to provide a set of quantified methods to determine the relations among the three types of intersection delay and how they each contribute to the total delay.

Intersection Delay Models

Unsaturated Straight and Left-Turn Delay Models
All analyses in this paper are based on the right-driving system unless otherwise stated.

Under unsaturated conditions, intersection delays are primarily caused by traffic control [23,24]. This paper first examines the delays from one approach of a typical crossing (four-legged, two-way) intersection, which has exclusive left-turn phases, and left-turning and through approaching traffic do not interfere with each other. Delays for both types of approaching traffic can be calculated [25].

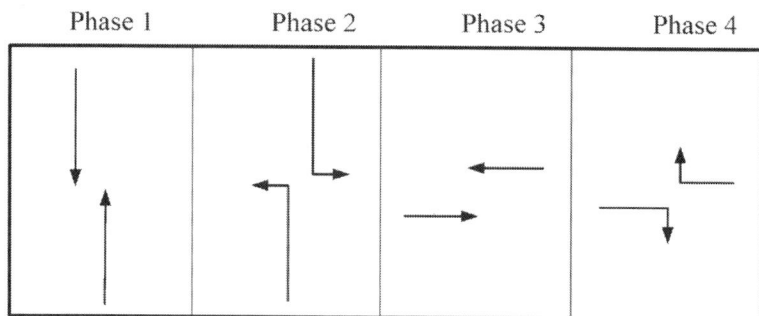

Figure 1. Phase diagram of a typical four-phase signal control.

The signal phases of a typical four-legged, two-way intersection with exclusive left-turn phases are illustrated in Figure 1. When the vehicle arrival rate and the capacity of an approach are fixed in any given signal cycle, there is a linear relationship between delay and arrival rate. In Figure 2, the x-axis shows the time t (s); the y-axis represents the total number of vehicles (veh); q is the vehicle arrival rate (veh/s); s is the saturated flow rate of the approach (veh/s); and c, r and g are cycle length, red and green intervals, respectively. The projection of ΔOAB on the x-axis is the delay of every vehicle arriving at the stop line (more precisely, the tail of the vehicle queue); and the projection on the y-axis is the number of vehicles queuing behind the stop line at different moments.

Therefore, in any given signal cycle, the phasing delay D (veh/s) of all vehicles coming into the approach equals the area of ΔOAB, which is:

$$D = S_{\Delta OAB} = \frac{1}{2} OA \cdot BD$$

(1)

where,

OA=r,BD=AD·tanα=AD·s,

also because,

$$AD = \frac{qr}{s-q}, \ [AD \cdot s = (r + AD) \cdot q],$$

(2)

therefore,

$$D = \frac{1}{2} OA \cdot BD = \frac{qsr^2}{2(s-q)}.$$

(3)

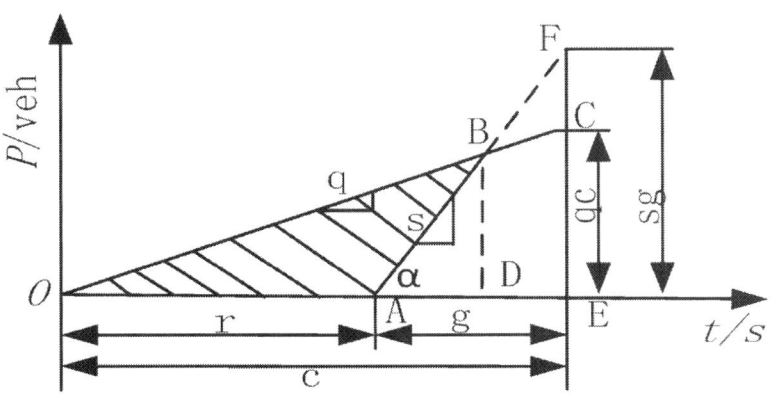

Figure 2. Unsaturated vehicle delay.

Right-Turn Delay Models

The delay of right-turning vehicles passing through crossing pedestrians can be modeled as the following process: in Figure 3, at S_0, the vehicle arrives at the pedestrian stream and decides whether there is a passable gap, i.e., whether there is a long enough time interval for it to pass through the pedestrian stream. S is the time when there is a passable gap for the vehicle to drive through the conflict area. Therefore, the delay can be calculated with $T = S - S_0$, with which we could, in turn, conclude that right-turn delay is primarily determined by three factors, namely:

1. the randomness of arrival of the right-turning vehicle;
2. the randomness of pedestrian arrival;
3. the minimum passing interval.

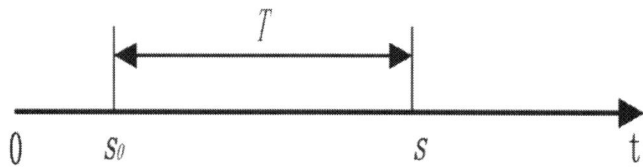

Figure 3. Right-turn delay.

Pedestrian crossing behavior can be modeled in a multi-row format as illustrated in Figure 4. Pedestrian crossing follows the Poisson distribution, and the interval between two adjacent rows obeys the exponential distribution. Minimal passing interval and pedestrian flow both follow the same approximate Poisson distribution.

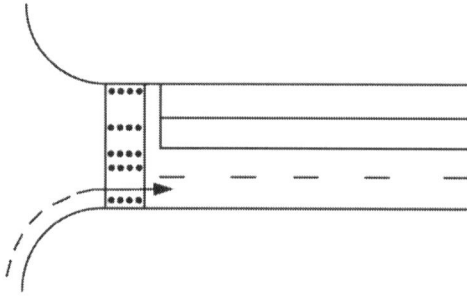

Figure 4. Illustration of the pedestrian crossing model.

Pedestrian crossing happens during the green interval. The process can be described in the following three steps (Figure 5): First, at the onset of the green signal, multiple rows of pedestrians with minimum intervals in between each row initiate the crossing process, which creates no passable gap for the vehicle. Next, pedestrian crossing starts to follow a Poisson distribution, and there are passable gaps for the vehicle to cross the conflict area. Finally, near the end of the green interval, vehicles can pass the crosswalk at will. u, v and w denote the time intervals of each step, respectively.

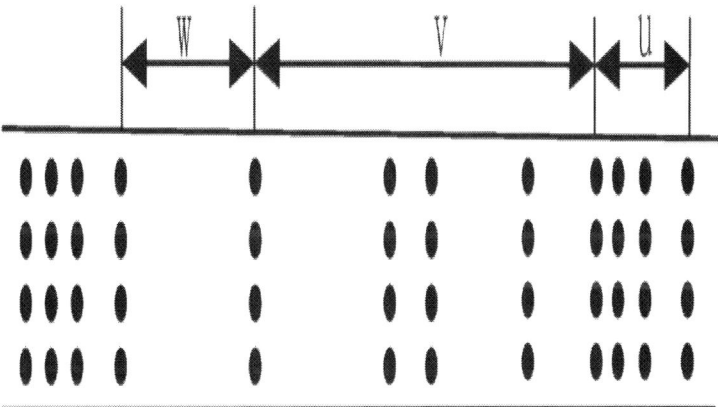

Figure 5. Illustration of pedestrian crossing process model.

Two types of right-turn delay models can be established following this pedestrian crossing process.

Type 1 Delay Model

The Type 1 delay model deals with delay caused by the random pedestrian stream, as illustrated in Figure 6. Variables are defined as follows: random variable θ is the time at which the vehicle arrives at the conflict area. Its value is the interval between the end of pedestrian crossing and the vehicle arrival time. The probability density of θ is $g(\theta)$, and the corresponding probability distribution is $G(\theta)$. The uniform distribution is chosen to describe the arrival time. Random variable Tt is the vehicle delay. In situations of Type 1 delay, 0≪Tt≪v. The probability density of Tt is ft (Tt), and the corresponding probability distribution is Ft (Tt).

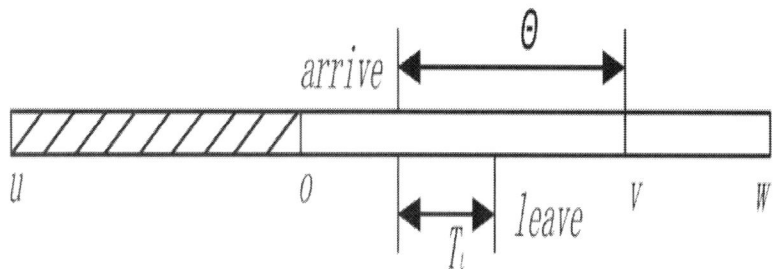

Figure 6. Illustration of the Type 1 delay model.

The vehicle arrives at θ moment, waiting for a passable gap. When the first passable gap appears, the vehicle drives through the conflict area, resulting in a delay of Tt. According to the remaining life theorem of Poisson streams and the assumption that passable intervals follow a Poisson distribution, we can calculate the remaining life of the passable interval, and the delay obeys the exponential distribution, i.e.:

$$F_t(T_t) = P\{T_t \le t\} = 1 - e^{-\lambda t}$$

(4)

where λ is the interval parameter and t is the time, t≫0.

The probability density of Tt is ft (Tt)=Ft′(Tt), therefore:

$$f^t(T^t) = \gamma e_{-\gamma t}$$

(5)

When the vehicle arrives at θ moment, its delay distribution function is $F_t(T_t,\theta)$, and the corresponding probability density is $f_t(T_t,\theta)$. When $0 \ll T_t < \theta$, right-turning vehicles can pass through immediately upon discovering a passable gap, and the distribution function becomes $F_t(T_t,\theta) = F_t(T_t)$. When $T_t \gg \theta$, vehicles can only pass through the conflict area after the crossing pedestrians, in which case $T_t = \theta$, and the distribution function becomes $F_t(T_t,\theta) = 1 - F(\theta)$, where $F(\theta)$ is the distribution function of θ. Therefore, the distribution function of T_t is:

$$F_t\left(T_t,\theta\right) = \begin{cases} 1 - e^{-\lambda t}, & 0 \leq t \leq \theta \\ e^{-\lambda\theta}, & t \geq \theta \end{cases}$$

(6)

The time interval for random crossing pedestrians is $[0,v]$. During $[0,v]$, the average delay D1 when the vehicle arrives at θ moment is:

$$D_1 = \iint t f_t\left(T_t,\theta\right) g\left(\theta\right) dt d\theta$$

(7)

With (5) and (6), we have,

$$D_1 = \frac{\lambda v + e^{-\lambda v} - 1}{\lambda^2 v}$$

(8)

$$\lambda = q e^{-q\alpha / 3600} \Big/ 3600$$

(9)

Where q is the pedestrian flow (p/h) and α is the time interval between pedestrian rows (s).

Type 2 Delay Model

The Type 2 delay is comprised of two phases. [0,u] is given as the time interval for multiple rows of pedestrians with no passable gaps in between them to leave the conflict area. As shown in Figure 7, the Type 2 delay can be represented by T'=t1+t2, where t1 is a random variable (0≪t1≪u), whose distribution function is F1(t1) and the corresponding probability density f1(t1). θ is the difference between the time pedestrians leave the conflict area and the arrival time of the right-turning vehicle.

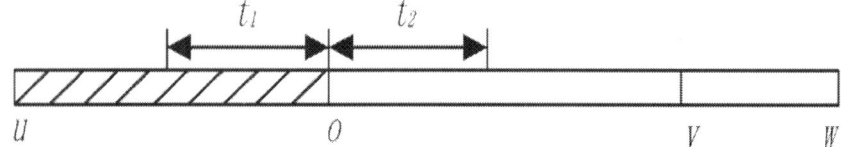

Figure 7. Illustration of the Type 2 delay.

When t1 and θ are both treated as uniformly distributed, the probability density of t1 is:

$$f_1 (t_1) = \frac{1}{u}$$

(10)

t2 is a random variable (0≪t2≪v), whose distribution function is similar to Equation (3), in which case, the vehicle arrival time is no longer a random variable within [0, v], but is a constant, i.e., the value of v. Therefore, the distribution function of t2 is:

$$F_2 (t_2, v) = F_t (t_2, v)$$

(11)

The average Type 2 delay D2 can be calculated with the following equations:

$$D_2 = E (T') = E (t_1) + E (t_2)$$

(12)

$$E\left(t_1\right) = \int_0^u t_1 f_1\left(t_1\right) dt_1$$

(14)

$$E\left(t_2\right) = \int_0^v t_2 f_2\left(t_2\right) dt_2 + v\left[1 - F\left(v\right)\right]$$

(14)

where f2(t2) is the probability density of t2.

By bringing (10), (12) and (13) into (14), we can have:

$$D_2 = \frac{u}{2} + \frac{1 - e^{-\lambda v} - \lambda v e^{-\lambda v}}{\lambda^2 v} + v e^{-\lambda v}.$$

(15)

When the arrival time is uniformly distributed, the average delay of right-turning vehicles is:

$$D = \frac{\left(u + v\right) D_2 + v D_1}{u + 2v}$$

(16)

To test the validity of the model, u, v and α are given the values of 10 s, 15 s and 5 s respectively. Right-turn average delays are calculated under conditions of five sets of pedestrian volume (q), i.e., 700 p/h, 800 p/h, 900 p/h, 1000 p/h and 1100 p/h. All of the values are then fed into Vissim, which is a computer-aided traffic simulation software. The results of the calculations and simulations are then compared.

In Table 1, average delays calculated with the model are either the same as or very close to the results of computer simulation, which suggests that when pedestrian volume is between 700 and 1100 p/h, the model can be used for estimating average right-turn traffic delay to a high degree of accuracy.

Table 1. Comparison of average delays.

Pedestrian Volume (p/h)	700	800	900	1000	1100
Simulation results (s)	10.1	10.1	10.2	10.4	10.8
Calculation results (s)	10.1	10.1	10.2	10.3	10.6

STRUCTURED DELAY ANALYSIS BASED ON SHIFT-SHARE ANALYSIS

Shift-share analysis has been widely applied in fields of regional science, political economy and urban studies to determine what portions of regional economic growth or decline can be attributed to national, industry and regional factors. A shift-share analysis attempts to identify the sources of regional economic changes, which helps identify industries in which a regional economy has competitive advantages over the larger economy. It is one of the major tools for regional economic and industrial structure analysis. This papers attempts to broaden its application to intersection delay analysis.

The Premise

With shift-share analysis, a change in regional traffic delay is seen as a dynamic process. Regional traffic or intersection delay is treated as the whole, whose change within a certain period of time is then subdivided into three components, which are the share effect, structural deviation effect and competitiveness effect.

SSA is used here to evaluate the structure of delay at an intersection through the analysis of the competitiveness of each component, with the purpose of determining the structure of future delay growth and, based on the result of which, to provide a more valid and scientific basis for future delay counter-measures.

The Modeling

The way SSA is applied to intersection delay analysis is by breaking apart the delay growth of the entire intersection into its components, i.e., first into the approaches of the intersection and, after that, further into the flows of each approach, namely through, left-turn and right-turn traffics.

The growth (G) in traffic delay at an intersection can be divided into three parts: (1) share effect (RS); (2) structural deviation effect (PS); and (3) competitiveness effect (DS). Delay growth is the sum of share effect, structural deviation effect and competitiveness effect, i.e., $G = RS + PS + DS$. What follows is how to calculate each effect.

(1) Share effect (RS):

$$RS = \sum_i Y_i^0 R \tag{17}$$

where i denotes left-turning, through or right-turning traffic flow; RS represents the delay growth of an approach that is proportionate to the total growth of the entire intersection; Y0i is the baseline delay of the flow of traffic that is represented by I; and R is the growth rate of the entire intersection delay. If RS is greater than real delay values, then the deviation is positive, otherwise negative.

(2) Structural deviation effect (PS):

$$PS = \sum_i Y_i^0 R_i - \sum_i Y_i^0 R = \sum_i Y_i^0 (R_i - R) \tag{18}$$

where PS is the difference of the delay growth between that calculated with the growth rate of a traffic flow and that calculated with the growth rate of the total delay. It reflects the relation between the growth rate of one traffic flow and the entire intersection. Additionally, Ri is the delay growth rate of the traffic flow, which is denoted by i.

(3) Competitive effect (DS):

$$DS = \sum_i Y_i^0 r_i - \sum_i Y_i^0 R_i = \sum_i Y_i^0 (r_i - R_i) \tag{19}$$

where DS is the difference of the delay growth between the actual growth from the traffic flow that is denoted by i and its calculated growth based on the total growth rate of the entire intersection. It reflects the traffic flow's "competitiveness" in contributing to the total delay growth of the intersection. ri is the actual delay growth from the traffic flow that is denoted by i.

Putting everything together, we have:

$$G = RS + PS + DS = \sum Y_i^0 R + \sum Y_i^0 (R_i - R) + \sum Y_i^0 (r_i - R_i)$$
(20)

CASE STUDY

The intersection between Yong-Chang road and East Bao-xiu road in the Chinese city of Kunming was picked as the site of application after careful examination of its traffic volumes and signal phasing design to ensure that it was indeed a typical four-phase intersection. Two seasons of delay data at the intersection were collected and gathered in Table 2:

Table 2. Delay statistics at the application intersection.

Approaches	Flow	1st Season	4th Season	Increase	Growth (%)
	Total	18	30	12	67
	Through	8	15	7	88
West-east	Left-turn	6	9	3	50
	Right-turn	4	6	2	50
	Total	23	41	18	78
	Through	13	22	9	69
North-south	Left-turn	7	13	6	86
	Right-turn	3	6	3	100

Here, we take the north approach as the subject of study, whose average left-turn, through and right-turn delays were calculated with the previously introduced formulas and gathered in Table 3.

Table 3. North approach delay data.

Traffic Flow	RS			PS		DS	
	Growth	Growth rate (%)	Difference	Growth	Growth rate (%)	Growth	Growth rate (%)
Through	8.67	67	1.67	2.71	21	-2.38	-18
Left-turn	4.67	67	1.67	-1.17	-17	2.5	36
Right-turn	2	67	0	-0.5	-17	1.5	50

From the above two tables, we could know:

1) All three traffic flows of the north approach, i.e., left-turn, through and right-turn, have higher delay growth rates than the delay growth of the entire intersection, which suggests that the north-south approaches have larger shares of causing the traffic delay at the intersection than the other direction. There is a newly finished business complex project to the northeast of the intersection, which could be the source of delay growth in the north-south direction.

2) Through traffic contributes much greater delay to the total delay than left-turn and right-turn traffics, which indicates that the largest share of delay for the intersection comes from through traffic. The increase of through volume and inefficient signal phasing design could be two of the main reasons.

3) Left-turn and right-turn delays from this approach appear to have higher growth rates than those of the other approaches. However, the through traffic contributes far less to the total delay growth than those of the other approaches. All of this indicates that the cause for the growth could be the traffic attraction from the newly finished business complex, and there is an emerging problem with the signal phasing design that needs to be adjusted to the changes of traffic volume.

CONCLUSIONS

Two areas of traffic delay study that demand more emphasis are right-turn delay and investigations into the effects of different types of delay on the intersection's total delay. This paper made an effort to address both of these issues. First, the process or mechanism of right-turn conflict was analyzed and discussed, followed by the establishment right-turn delay models. Second, shift-share analysis was introduced to help provide a structured method of analyzing how each type of traffic delay affects or contributes to the total delay of the intersection. This new approach provides a quantified method of examining the delay changes of different traffic flows within an intersection, their relations and how they contribute to the total delay change of the intersection. The models and methods in this paper can help broaden the options in intersection delay study and provide more scientific bases for intersection planning, lane management, channelization and signal phasing design, etc.

Conflicts between left-turning vehicles and pedestrians in the real-world can be rather random and hard to describe accurately with models; therefore, the right-turn delay models in this paper can only be used as an attempt at calling for more emphasis on related studies and as a reference point to other researchers. The use of shift-share analysis in this paper is still rudimentary and lacks more tests with real-world data, but its potential in studying the relations between different delays and the total intersection delay is well worth further looking into.

What is worth noticing is that the method and model are only applicable when there is a dedicated right-turn lane. When the queue at the right-turning lane overflows or there is a mixed through and right-turning lane existing, the proposed model and the method would be very different from what has been discussed in the paper. Furthermore, all of the analyses were based on right-hand driving situations. Whether the model could be used for the study of left-turn delay in left-hand driving situations could be a topic for future studies

ACKNOWLEDGMENTS

The author would like to thank the referees for their helpful comments.

AUTHOR CONTRIBUTIONS

Jianfeng Xi wrote this paper. Wei Li, Shengli Wang and Chuanjiu Wang were involved in all research activities and provided many valuable suggestions. All authors discussed and contributed to the manuscript. All authors have read and approved the final manuscript.

REFERENCES

1. Kelley, K.P.; Martin, T.; Paul, P. Evaluation of average delay as a measure of effectiveness for signalized intersections. Transp. Res. Board **2001**, 12, 398–453.
2. Chen, S.; Guo, J.; Wang, X. Analysis and Simulation on Signalized Intersection Delay. J. Beijing Jiaotong Univ. **2005**, 3, 77–80.
3. Mousa, R.M. Analysis and modeling of measured delays at isolated signalized intersection. J. Transp. Eng. **2002**, 18, 347–354.
4. Akcelik, R.; Rouphail, N.M. Overflow Queues and Delays with Random and Platooned Arrivals at Signalized Intersections. J. Adv. Transp. **1997**, 28, 227–251.
5. Akcelik, R. The Highway Capacity Manual Delay Formula for Signalized Intersections. ITE J.**1988**, 58, 23–27.
6. Reid, D.H. Delays Caused by Right-Turning Vehicles. Transp. Sci. **1968**, 2, 160–171.
7. Hagita, K. Right-turning vehicle capacity at a signalized intersection of a multilane road. Infrastruct. Plan. Rev. **2001**, 18, 949–956.
8. Rahim, F.; Yoassry, M. Multi-regime arrival rate uniform delay models for signalized intersections. Transp. Res. A **2001**, 35, 625–667.
9. Liu, G.; Pei, Y. Study of Calculation Method of Intersection Delay under Signal Control. China J. Highway Transport **2005**, 1, 104–108.
10. Yaqub, O.; Li, L.; Noureldin, A. Modeling and Analysis of Connected Traffic Intersections Based on Modified Binary Petri Nets. Int. J. Veh. Technol. **2013**, 2013.
11. Ni, Y.; Li, K. Conflict of Right-turn Vehicles and Pedestrian at Signal Intersection. Comput. Commun. **2007**, 25, 22–26.
12. Su, Y.; Yao, D.; Zhang, Y. Exclusive right-turn phase setting simulation based on vehicle/pedestrian conflict analysis. J. Highway Transp. Res. Dev. **2009**, 26, 133–138.
13. Zhang, M.; Liu, Y.; Pei, Y. Control of right-turn vehicles at signalized intersections based on comfort and safety of pedestrian. J. Highway Transp. Res. Dev. **2008**, 25, 136–139.
14. Qureshi, M.A.; Han, L.D. Delay model for right-turn lanes at signalized intersections with uniform arrivals and right turns on red. Transp. Res. Record **2001**, 1776, 143–150.
15. Guo, X.; Dunne, M.C.; Black, J.A. Modeling of pedestrian delays with pulsed vehicular traffic flow. Transp. Sci. **2004**, 38, 86–96.

16. Zhou, Z.; Wang, W.; Ren, G.; Gong, X. Choice model of pedestrian crossing behavior at signalized intersections. J. Southeast Univ. **2013**, 43, 664–668.
17. Tao, P.; Wang, D.; Jin, S. Mixed distribution model of vehicle headway. J. Southwest Jiaotong Univ. **2011**, 46, 633–637.
18. Zhao, J.; Bai, Y.; Yang, X. A regression-based pedestrian-bicycle adjustment factor for capacity of right-turn movements. J. Highway Transp. Res. Dev. **2012**, 29, 120–126.
19. Casler, S.D. A Theoretical Context for Shift and Share Analysis. Reg. Stud. **1989**, 23, 43–48.
20. Rafels, C.; Vilella, C. Proportional share analysis. TOP **2007**, 15, 341–354.
21. Dinc, M.; Haynes, K.E. International Trade and Shift-Share Analysis: A Specification Note. Econ. Dev. Q. **1998**, 12, 337–343.
22. Mehregan, N.; Asgary, A.; Rezaei, R. Effects of the Bam earthquake on employment: A shift-share analysis. Disasters **2011**, 36, 420–438. [PubMed]
23. Francis, D.; Hesham, R.; Kang, Y.-S. Comparison of delay estimates at under-saturated and over-saturated pre-timed signalized intersections. Transp. Res. B **2004**, 38, 99–122.
24. Harikishan, C. Over saturation inherence and traffic diversion effect at urban intersections through simulation. J. Transp. Syst. Eng. Inf. Technol. **2009**, 9, 72–82.
25. Kim, J.-T. Improvement on estimation of the uniform delay of an exclusive permitted left- turn lane group. J. Transp. Eng. **2006**, 132, 708–714.

CITATION

Jianfeng Xi, Wei Li, Shengli Wang,and Chuanjiu Wang, An Approach to an Intersection Traffic Delay Study Based on Shift-Share Analysis, doi:10.3390/info6020246.

CHAPTER 8

Using Multi-Attribute Decision Factors for a Modified All-or-Nothing Traffic Assignment

EunSu Lee [1,*] and Peter G. Oduor [2]

[1]Upper Great Plains Transportation Institute, North Dakota State University, Fargo, ND 58105, USA

[2]Department of Geosciences, North Dakota State University, Fargo, ND 58105, USA; E-Mail: peter.oduor@ndsu.edu

ABSTRACT

To elucidate a realistic traffic assignment scenario, a multi-criterion decision system is essential. A traffic assignment model designed to simulate real-life situation may therefore utilize absolute and/or relative impedance. Ideally, the decision-making process should identify a set of traffic impedances (factors working against the smooth flow of traffic) along with pertinent parameters in order for the decision system to select the most optimal or the least-impeded route. In this study, we developed geospatial algorithms that consider multiple impedances. The impedances utilized in this study included, traffic patterns, capacity and congestion. The attributes of the decision-making process also prioritize multi-traffic scenarios by adopting first-in-first-out prioritization method. We also further subdivided classical impedance into either relative impedance or absolute impedance. The main advantage of this innovative multi-attribute, impedance-based trip assignment model is that it can be implemented in a manner of algebraic approach to utilize shortest path algorithm embedded in a Geographic Information Systems (GIS)—Graphical User Interface tool. Thus, the GIS package can therefore handle the multi-attribute impedance effectively.

Furthermore, the method utilized in this paper displays flexibility and better adaptation to a multi-modal transportation system. Transportation, logistics, and random events, such as terrorism, can be easily analyzed with pertinent impedance.

INTRODUCTION

A traffic assignment model (TAM) employs various approaches in issues-of-scale with respect to modal transportation networks. Since transportation systems play an important role in the movement of goods and people, a TAM must be contextually and scientifically sound. User equilibrium invokes Wardrop's first principle [1], which can be surmised as: In equilibria, there is no unilateral gain for a driver to reduce his/her travel costs by selecting another feasible route. This can be simply regarded as the journey-time equilibrium for route choice, where travel cost on routes selected are set as equal but less than on any unused route [1]. In a typical user equilibrium scenario, a user may choose the minimum cost route. However, since the user equilibrium can vary due to random perceptive components, such as user preference, experience, and time of travel, it is necessary to include other routes that may meet a set threshold. To account for this randomness, stochastic user equilibrium assignment can be utilized since it incorporates unilateral route changes. Under the stochastic user equilibrium assignment scenario, various assumptions can be stated: (a) users cannot reduce the travel cost by changing routes owing to limited available information at that particular instant; and (b) congestion can be adequately accounted for by using a delay function [2,3,4,5]. As such, stochastic user equilibrium becomes a fixed-point solution to the problem of selecting routes and travel times over a network. User choices are made according to a random utility model for a stochastic user equilibrium model, and link travel times are dependent on the link flows.

One of the traffic assignment models for a large-scale travel demand model is a path-based all-or-nothing (AON) assignment using Dijkstra's shortest path algorithm [6]. The AON assignment allocates all vehicle trips to an origin and destination pair, which is the most feasible single route. There are advantages of this approach; the foremost is the anthropogenic-induced (social optimum) factor. Social optimum status (macroscopic characteristic) is commonly used for large-scale network transportation planning [7,8], where maximal profits can be derived at minimized risks.

However, this is quite unlike for short-term trip estimation in the user equilibrium, which can indeed take into account travel-time dependency (microscopic characteristic) by considering both origin-destination flow and capacity [5]. An aggregated origin-destination flow can be disaggregated into individual trips to reflect travel-time dependency. The disaggregated trips would thereafter be assigned systematically to display an ideal route in stochastic user equilibrium. The drawback in this case would be the large number of links required to formulate contiguous links. Thus, most heuristic algorithms that would employ such an aggregation process in route assignment should not exceed a set of impedance thresholds especially on number of route links and their connectivity properties [9].

In light of existing traffic assignment models, the TAM does not feature pertinent link segment information, such as road closure, underlying or overarching ad hoc policy (e.g., converting a three-lane road to a two-lane road with a bicycle route), user route preference based on experience, and vehicular performance related factors (e.g., driving a fully-loaded truck uphill versus using a circuitous route), we are proposing a GIS-based traffic assignment model that can fully account for these factors and more in efficient ways without relying on transportation planning software on the shelf. Our basic principle is the use of impedance as the elemental guiding paradigm of the developed traffic assign model, which can easily be extended or integrated into any applied GIS network model with little extra efforts and costs. This would offer: (i) flexibility in vehicle routing; and (ii) easiness of incorporating an attribute data structure that would account for all impedance cases. The proposed modified all-or-nothing (MAON) assignment place several packets of a bundle of the vehicle trips to an origin and destination pair, which is the most feasible single route for each packet at a time.

The major objectives of this study are to develop a multi-attribute and impedance-based traffic assignment model for transportation planners and travel demand analysts. We then applied the proposed traffic assignment model to simulate an urban traffic scenario incorporating six major impedance attributes.

LITERATURE REVIEW

Selecting a route process typically involves minimizing travel distance, especially under the assumption that transportation cost (also referred to as disutility) is significantly affected by travel distance. Janic showed that a complete logistical cost should entail an economic distance [10]. An economic distance is the distance a commodity may travel before transport costs exceed the value of freight. A traffic assignment model involving a travel distance minimization problem include Beyer's line weight means (LWM) [11]. Dijkstra's shortest path algorithm is easily utilized in LWM as a means of determining a route with the least impedance. The disadvantages of utilizing LWM include: (i) the fact that only single-modal networks can be modeled; (ii) segment gaps may be inadvertently included in the route selected; and (iii) total impedance can be too small to interpret for very short segments. Nevertheless, LWM is still useful in the long-range transportation planning, especially in comparing economic distances between newer routes with other alternative routes.

Operational disutility measures, such as, travel time, congestion, and operating speed, have been used to gauge impedance [12]. Most studies use travel time as the dominant impedance factor. It has been postulated that viable future simulation approaches would require including faster travel time paths and segments with increased travel speed as path alternatives [13,14,15,16].

Nesterov and de Palma's approach constrained the flow capacity on a single link for simulating congestion by considering volume and capacity [17]. The rationale of their approach stemmed from setting both demand and capacity constraints over a finite time frame [18]. The traffic reassignment, in such a scenario, consists of two processes: (a) deducting the traffic flow leaving the congested segment from a prior time, which we can term as an apriori condition; and (b) adding the traffic entering the congested segment during a current period as a posteriori condition [19]. In this way, a sequence of trips assigned was designed by a first-in first-out (FIFO) queuing algorithm. The FIFO rule within the algorithm is required for multiple realistic origin-destination assignments [20,21]. This FIFO rule can also be applied to this study in order to a similar approach to a dynamic stochastic user equilibrium-based traffic assignment model by disaggregating the selected routes into multiple routes using a random-order sequence over a specified period.

Geographic information systems (GIS) can configure solutions for re-routing problems, for example, in real-time emergency vehicle routing when a user has to consider physical dynamics (e.g., one way roads, traffic, and road maintenance) and inherent uncertainty (e.g., parade routes, pedestrian traffic, etc.). The classical GIS approach is to set single impedance and then aggregate all the segments that meet the impedance condition. To implement sophisticated network analyses and incorporate geospatial information, finer levels of impedances that may vary from segment to segment are required. For instance, Kwan and Ransberger applied segment-level impedance designated to model where impasses were most likely to exist [22]. They also noted that dealing with a degree of uncertainty was cumbersome, especially when factors like debris on road segments after storms impede travel on the roads. To account for multiples factors impeding travel, multi-criteria decision-making process have been utilized in operations research in general [23,24].

Grossardt et al. [23] used an analytical hierarchy process in transport. Chen et al. [24] used an analytic minimum impedance surface (AMIS) model on contiguous surfaces for identifying route choice based on input from a user. The analytical hierarchy process can also be extended to include a compensatory multi-criteria decision method and include risk analysis [25].

In précis, a comprehensive study addressing a multi-attribute impedance approach with a first-in first-out rule within thresholds of the packet size has never been undertaken. We believe that this is the first algorithm that does quantify, annotate and provide a visual output of traffic assignment model that addresses: volume/capacity ratio, driving hazard conditions (flooded sections), bridge closures, general road repair/construction, right-of-way, road classification (unpaved or otherwise) together. The model proposed in this study is similiar to a dynamic traffic assignment model, which can be utilized for a state and regional long-range transportation planning. The model used for this study disaggregates the period into smaller time intervals, such as, weekly or monthly.

MODEL DEVELOPMENT

Approach

A network dataset consists of nodes and links. Nodes represent: (i) locations; (ii) terminating nodes; and (iii) centroids of areas Links are road segments with directions and capacity. Each link is connected by two nodes (a starting node and an ending node) with the end node serving as the starting node for the next contiguous link. Each segment contains associated generalized costs such as distance and travel time. A route is a set of segments connected from origin to destination through multiple segments and have impedance inevitably. The impedance is a sum of generalized costs from the segments, which belong to the route. For example, if a generalized cost of the segments is travel time, the impedance of the route is travel time, and the best route is the quickest route. A detour, with the least impedance value, might be selected when one of the segments of the route increases the generalized cost abruptly. A trip can be defined as the journey between an origin and destination through a route. An origin and destination pair may likewise include node information representing centroids of areas and points.

Secondarily, selected routes generate a dataset of impedance depending on geographic locations, types of movement, transport modes, and time windows. The aim in such an optimized system is minimizing the impedance resulting from temporal and spatial gaps. The impedance is sourced from transportation infrastructure, weather, policies and regulations, vehicles, commodities, and drivers. The impedance on a logical transportation networks appears onto nodal and link components, as modified from Gutiérrez and Urbano [26]:

$$Z_p = \sum_{s \in p} Z_s + \sum_{s \in p} Z_n \tag{1}$$

where Z_p = impedance (i.e., total generalized costs of links) of a path p; Z_s = generalized cost of link s along the feasible path p; Z_n = generalized cost of a node n along the feasible path p.

If an activity reaches adjacent nodes, the connected segments become alternative segments of a route, that is, p = (S, N), of which movement from origin to destination through a set of links can be represented by S(i, j). S is a set of links, and N is a set of nodes including i and j. Node i is the tail node (start node), and node j is the head node (end node).

Components

We expanded Arnold et al.'s [27] "open" or "closed" electric circuit concept for assessing disconnection and connection from a tail node i to a head node j. When goods and people travel from node i to j via contiguous segments, the status of the connecting segment is expressed by a Boolean variable $X_s = 1$, which implies that the segment s is traversable due to its high traversability (i.e., low impedance (In electronics, zero means that an electric circuit is open, so electrical current cannot flow: The analogy in transportation implies that a road is closed due to a high impedance value)). In this study, absolute traversability, X, is binary values, which represent the maximum traversable value of 1 and the minimum traversable value of 0. We define that the reverse concept of traversability is impedance by using absolute complements, which is denoted by I. For example, if impedance is high, the traversability is low and vice versa. In other words, high absolute impedance on a link (I_s), in this study, was defined as the low absolute traversability on the link (X_s).

Examples of absolute traversability on a link (X_s) included increasing number of lanes and fair weather conditions. Absolute traversability on a node (X_n) is found from a normal condition of a bridge. On the contrary, one way highway can improve a traffic flow for forward direction, but prevent contraflow (Table 1). In the same way, bad weather condition impedes travel along a road segment. Examples of node impedance can be found from bridges, terminals, and border crossings. A bridge is either opened or closed. Examples of absolute impedance on a link (I_s) were attributed to one-way streets for Cases 5–8 and inclement weather for Cases 2, 4, 6, and 8 (Table 1). Similarly, absolute impedance on a node (I_N) can take the form of an additional impediment, such as presence of transshipment between two modes and park-and-ride facilities. Impedance was adapted as a unitless value that is normalized, resulting in a relative impedance value weighted by segment length. The nodes should be transformed into dummy segments with generalized, virtual costs in GIS, so the points are converted into line shapefiles. For that reason, this study considers the nodes as segments for the rest of this paper.

Absolute traversability X_s was simplified by an AND logic representing logical conjunctions (Table 1). To be traversable through a

AND logic, all attributes (k) should provide traversability. X_{sk} represents generalized cost of an attribute k on a segment s. If a factor k impedes travel, the segment s is declared untraversable (non-traversable). When the impedance is extremely high on a segment s due to one or more of k attributes as absolute barrier, the segment can become untraversable (non-traversable), for example, when $X_s = 0$ in the relation:

$$X_s = \mathrm{AND}(X_{sk})$$

(2)

In this way, the AND logic will contain eight combinations (i.e., 2^3) in total (Table 2). From Equation (2) and Table 1, if at least one of the three attributes of a segment ($X_{s,k=\{1,2,3\}}$) shows un-traversability, the segment s (X_s) would be untraversable.

Table 1. Examples of *AND* logic on absolute impedance values of a segment s.

| Case | One-Way $(X_{sk=1})$ | Bridge $(X_{sk=2})$ | Weather $(X_{sk=3})$ | Traversability | | | Impedance | |
				AND (X_{sk})	Absolute Traversability (X_s)	1-AND (X_{sk})	Absolute Impedance (I_s)
Case 1	1	1	1	1	Traversable	0	No Impedance
Case 2	1	1	0	0	Untraversable	1	Very High
Case 3	1	0	1	0	Untraversable	1	Very High
Case 4	1	0	0	0	Untraversable	1	Very High
Case 5	0	1	1	0	Untraversable	1	Very High
Case 6	0	1	0	0	Untraversable	1	Very High
Case 7	0	0	1	0	Untraversable	1	Very High
Case 8	0	0	0	0	Untraversable	1	Very High

Table 2. Determining relative impedance value of a segment s.

Case	One-Way ($R_{sk}=1$)	Bridge ($R_{sk}=2$)	Weather ($R_{sk}=3$)	Traversability $\prod_{k=1}^{3}P(R_{sk})$	Relative Traversability $P(R_s)$	Impedance $1-\prod_{k=1}^{3}P(R_{sk})$	Relative Impedance (I_s)
Case 1	1	1	1	1	Traversable	0	Very Low
Case 2	1	1	$P(R_{s,k=3})$	$P(R_{s,k=3})$	Most Likely	$1-P(R_{s,k=3})$	Low
Case 3	1	$P(R_{s,k=2})$	1	$P(R_{s,k=2})$	Most Likely	$1-P(R_{s,k=2})$	Low
Case 4	1	$P(R_{s,k=2})$	$P(R_{s,k=3})$	$P(R_{s,k=2})\times P(R_{s,k=3})$	Likely	$1-P(R_{s,k=2})\times P(R_{s,k=3})$	High
Case 5	$P(R_{s,k=1})$	1	1	$P(R_{s,k=1})$	Most Likely	$1-P(R_{s,k=1})$	Low
Case 6	$P(R_{s,k=1})$	1	$P(R_{s,k=3})$	$P(R_{s,k=1})\times P(R_{s,k=3})$	Likely	$1-P(R_{s,k=1})\times P(R_{s,k=3})$	High
Case 7	$P(R_{s,k=1})$	$P(R_{s,k=2})$	1	$P(R_{s,k=1})\times P(R_{s,k=2})$	Likely	$1-P(R_{s,k=1})\times P(R_{s,k=2})$	High
Case 8	$P(R_{s,k=1})$	$P(R_{s,k=2})$	$P(R_{s,k=3})$	$P(R_{s,k=1})\times P(R_{s,k=2})\times P(R_{s,k=3})$	Rarely Likely	$1-P(R_{s,k=1})\times P(R_{s,k=2})\times P(R_{s,k=3})$	Very High

Note: The possibility value of traversability R_{sk} lies between 0 and 1, that is, $P(R_{sk})=[0,1]$.

Some segments of a selected path can be classified as "weak segments" using a degree of traversability value, $P(R_s)$, defined by the distribution [0,1] such that $P(R_s)=[0,1]$. This algebraically means that it is relatively traversable, due to a weak link, which has a value of relative traversability within the range of $(0 < P(R_s) < 1)$ [28,29]. The relative traversability on a link $[P(R_s)]$ is the product of all associated traversability factors. By using this, an assignment process can be akin to road users opting to avoid traveling on a rough section of a route by choosing to go through a construction zone where they may be delayed further. Relative traversability also describes the ease of access to traversable routes [30,31]. Three attributes will generate eight combinations (i.e., 2^3) in total (Table 2). Relative traversability of a segment can therefore be obtained from:

$$R_s = \left(\prod_{k=1}^{K} P(R_{sk})\right) \forall k = \{1,2,\ldots K\}$$

(3)

where $\prod Kk=1P(Rsk)$ is the product of all $P(R_{sk})$, with k independent attributes of a segments. $P(R_{sk})$ represents relative traversability caused by an attribute k on a segment s, and $P(R_{nk})$ represents relative traversability caused by an attribute k on a node n (Table 2). The nodes are converted into lines in GIS, so the notations of the node n will be interchangeable with the notation of segment s. Because of the heterogeneous length of the segments, this study adopts weighted distance relative (WDR) impedance to reevaluate relative impedance (I) given by $WDR_s = d_s \times (1 - P(R_s))$, where $P(R_s)$ is relative traversability, and d_s is the length of segment s (e.g., [16]). By doing so, the sensitivity of the impedance of a segment increases to a traffic assignment model. The total impedance for a path P (i.e., Z_p) takes into account the total WDR for a route, that is, $Zp=\sum s \in PWDRs$, when the selected and preferred segments are integral to the path. Each component in Table 2 has a possibility value between 0 and 1. Once the length of a segment is applied to the route impedance, the Z_p is unitless. The output will also contain eight (2^3) possible combinations in total. Among them, only one case (Case 1) is absolutely traversable. Other combinations can be determined based on a most likelihood case.

CASE STUDY

The study area lies within the Fargo-Moorhead metropolitan area including the city of Prairie Rose. These cities were significantly affected by the spring flood of the Red River in March 2009 and March 2010. Several bridges were flooded, some were closed, and some had flood warning signs. Some regions along the Red River were placed under a flood watch. Two major interstate highways, I-29 and I-94, run perpendicular to each other. There is also a toll bridge between Fargo and Moorhead. During these floods, congestion was observed in routes that are usually passable year round. Some bridges on both sides of the river were partially submerged. In addition to areas close to the river, some lower regions were flooded by storm water, causing road lane closures.

System Optimum

For the first scenario, we assumed significant Red River flooding within flood-prone areas of Fargo-Moorhead. A variety of impedance attributes were set as travel retardants between the cities. If a journey has to follow rerouted trip, then the most likely path would be selected by the algorithm. A portion of the algorithm adopts Dijkstra's shortest path to minimize the total impedance from origin to destination only if a user were to select this as the most important criteria. To assign all trips to the feasible paths, the algorithm iterates through the path selection process at least P times, where P indicates the number of paths Equation (4). The number of iterations was estimated by optimizing the function:

$$minimize \sum_{p=1}^{P} Z_p = \sum_{p=1}^{P} \left(\sum_{s \in p} WDR_s \right)$$

(4)

If the number of trips and road capacity was unlimited, the algorithm needed only to run once to determine the preferable path based on an AON assignment.

Key Attributes

Let us suppose that travelers care about six major attributes namely: (a) volume/capacity (V/C) ratio; (b) hazards; (c) bridge operation; (d) construction; (e) right-of-way; and (f) road classification (Figure 1). Each attribute may include multiple levels, with unique possibility values ranging between 0 (high impedance) to 1 (no impedance). We can also safely assume: (i) non-compensatory methods (e.g., [25], pp. 621–680), which do not permit trade-offs between attributes; and (ii) also that the possibility produced by the attribute's levels follows a linear function. For simplification and ease of coding in GIS, we can also likewise assume that the volume capacity ratio follows a linear function (Figure 1a). As an example; if flood information is not available on a road segment and/or a bridge, the hazard possibility is set to 1, whereas the impedance within flooded segments is set to 0 (Figure 1b). Depending on the flood level, a bridge can be closed for safety concerns and algorithmically the segment could be set with a value of 0 indicating high impedance.

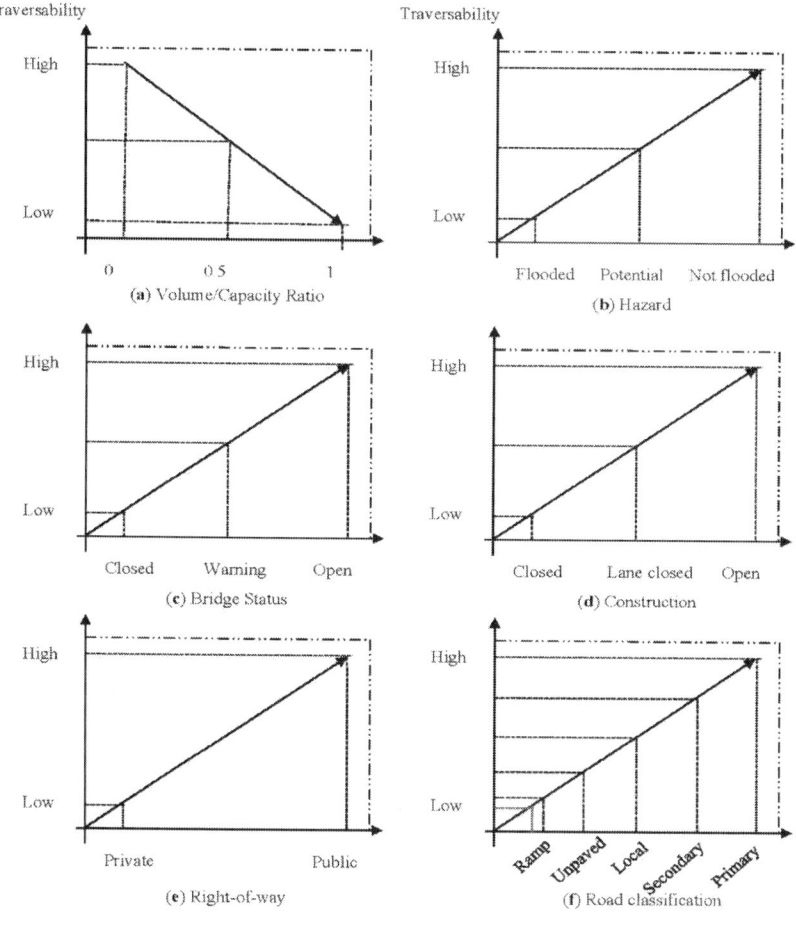

Figure 1. Levels of attributes and traversability. (**a**) Volume/capacity ratio; (**b**) hazard; (**c**) bridge status; (**d**) construction; (**e**) right-of-way; (**f**) road classification.

As an event level increases, a traveler will be aptly warned (Figure 1c). In urban settings, some roads may be closed due to road construction and therefore a high impedance value of 0 can be generated. If the road segment has multiple lanes, a partial lane or one or more lanes may be closed, limiting traffic to a narrow single lane (Figure 1d), an impedance value of 0.5 can be set based on the linearity assumption. In our approach, utilizing TIGER®(Topologically Integrated Geographic Encoding and Referencing) files, we segregated private from public roads. We set private road impedance at a high value (Figure 1e). TIGER® road files provide several road classifications, for example ramp, unpaved, local, secondary, and primary. We set impedance values as: 0.2 for ramp, 0.4 for unpaved, 0.6 for local, 0.8 for secondary roads, and no impedance for primary roads (Figure 1f).

Scenario Analyses

In the output tables, each route contains a unique route identifier representing a trip. The travel time (in hours) and the distance (in miles) were determined from the selected least impedance route. Attributes of the network dataset included the relative and absolute impedance values calculated for each segment using Equations (2) and (3), respectively.

- Scenario 1 (modified all-or-nothing (MAON) and trip set less than or equal to capacity): The first scenario shows capacity constraint with a MAON assignment for a trip set between origin and destination (Scenario 1 in Table 3). The number of trips was maximized at 900, a value lower than the average segment capacity specified, that is, 1060.

Table 3. Input information for Scenarios 1–5.

Origin		Destination		Trips (Packet Size)				
Town	FIPS	Town	FIPS	Scenario 1	Scenario 2	Scenario 3	Scenario 4	Scenario 5
Prairie Rose	64,320	Moorhead	43,864	900	1350	450	450	225
Prairie Rose	64,320	Moorhead	43,864			450	450	225
Prairie Rose	64,320	Moorhead	43,864			N/A	450	225
Prairie Rose	64,320	Moorhead	43,864					225
Prairie Rose	64,320	Moorhead	43,864					225
Prairie Rose	64,320	Moorhead	43,864					225
Prairie Rose	64,320	Fargo	25,700	900	1350	450	450	225
Prairie Rose	64,320	Fargo	25,700			450	450	225
Prairie Rose	64,320	Fargo	25,700			N/A	450	225
Prairie Rose	64,320	Fargo	25,700					225
Prairie Rose	64,320	Fargo	25,700					225
Prairie Rose	64,320	Fargo	25,700					225

Note: Place FIPS 64,320 is for Prairie Rose, ND, Place FIPS 43,864 for Moorhead, MN, and Place FIPS 25,700 for Fargo, ND.

- Scenario 2 (MAON and trip set larger than capacity): The second scenario involves a number of trips exceeding segment capacity. This was to demonstrate the route selection process under a limited set capacity.

- Scenario 3 (multiple trip sets when trip set is less than or equal to capacity): The third scenario has half the number of trips as in Scenario 1. It is also worthwhile to note that congestion effect is not easily discernible from Scenario 1.
- Scenario 4 (congestion effect and multiple trip sets): In this scenario, newer routes were selected between Prairie Rose and Moorhead. Scenario 4 is an extension of Scenarios 2 and 3, that is, using half of the trip set size from Scenario 2 to demonstrate congestion effect. The number of trips in the trip set was fewer than that from Scenario 2, although the total number of trips remained the same. We can deduce that for the OD pair, a distinct route will be selected independent of previously selected routes because of congestion and capacity limits.
- Scenario 5 (multiple trip sets coupled with random order of the trip sets): Randomly sequenced OD pairs were used to remove any sequence bias (Table 3). A trip set was sub-divided into two trip sets from Scenario 4. Scenario 5 illustrates the effect of a smaller trip set size. Scenario 5 also shows the relationship between impedance and miles generated since the distance information was an integral component of impedance.

RESULTS AND IMPLICATIONS

Effects of Capacity

In Scenario 1, the sequence of IDs (i.e., identifications of paths) indicates the order of the best routes based on the lowest impedance values for each pair (Table 4 and Figure 2a). The ID 0 is assigned for a trip set of 900 trips. The corresponding route length is also shown in the output table. Note that the travelled miles do not necessarily represent the shortest route distances. A packet of trip between Prairie Rose and Fargo was split into smaller size of trip packets due to a capacity of the path ID 1.

Scenario	ID	Origin	Destination	Trips	Impedance	Hour *	Miles **
1	0	Prairie Rose	Moorhead	900	9.93	0.17	8.28
	1	Prairie Rose	Fargo	160	6.24	0.14	5.39
	2	Prairie Rose	Fargo	740	6.40	0.16	5.10
		Average		600	7.52	0.16	6.26
2	0	Prairie Rose	Moorhead	1060	9.93	0.17	8.28
	1	Prairie Rose	Moorhead	290	10.92	0.25	8.35
	2	Prairie Rose	Fargo	770	6.76	0.16	5.18
	3	Prairie Rose	Fargo	290	9.21	0.20	7.57
	4	Prairie Rose	Fargo	210	9.72	0.19	7.51
	5	Prairie Rose	Fargo	80	10,006.69	0.15	6.27
		Average		450	1675.54	0.187	7.19
3	0	Prairie Rose	Moorhead	450	9.93	0.17	8.28
	1	Prairie Rose	Moorhead	450	10.35	0.19	8.38
	2	Prairie Rose	Fargo	450	6.31	0.16	5.06
	3	Prairie Rose	Fargo	160	6.72	0.14	5.41
	4	Prairie Rose	Fargo	290	7.00	0.17	5.43
		Average		360	8.06	0.17	6.51
4	0	Prairie Rose	Moorhead	450	9.93	0.17	8.28
	1	Prairie Rose	Moorhead	450	10.35	0.19	8.38
	2	Prairie Rose	Moorhead	160	10.75	0.22	7.98
	3	Prairie Rose	Moorhead	290	11.10	0.25	8.36
	4	Prairie Rose	Fargo	160	6.65	0.14	5.38
	5	Prairie Rose	Fargo	290	6.90	0.17	5.40
	6	Prairie Rose	Fargo	450	7.29	0.17	5.32
	7	Prairie Rose	Fargo	320	7.34	0.17	5.58
	8	Prairie Rose	Fargo	130	7.76	0.18	5.94
		Average		300	8.67	0.18	6.74

0	Prairie Rose	Moorhead	225	9.93	0.17	8.28
1	Prairie Rose	Fargo	225	6.14	0.14	5.39
2	Prairie Rose	Moorhead	225	10.34	0.23	8.00
3	Prairie Rose	Fargo	225	6.50	0.13	5.39
4	Prairie Rose	Moorhead	225	10.68	0.22	7.95
5	Prairie Rose	Fargo	225	6.62	0.14	5.46
6	Prairie Rose	Moorhead	225	10.99	0.23	8.11
7	Prairie Rose	Fargo	160	6.81	0.16	5.11
8	Prairie Rose	Fargo	65	7.10	0.15	5.49
9	Prairie Rose	Moorhead	160	11.36	0.22	8.09
10	Prairie Rose	Moorhead	65	11.53	0.19	8.98
11	Prairie Rose	Fargo	30	7.29	0.15	5.44
12	Prairie Rose	Fargo	195	7.42	0.16	5.19
13	Prairie Rose	Moorhead	95	11.96	0.25	8.94
14	Prairie Rose	Moorhead	30	12.10	0.24	8.46
15	Prairie Rose	Moorhead	100	12.95	0.24	9.69
16	Prairie Rose	Fargo	95	8.69	0.19	6.06
17	Prairie Rose	Fargo	60	8.84	0.18	6.23
18	Prairie Rose	Fargo	70	10.17	0.19	7.55
	Average		142	9.34	0.19	7.04

Note: * travel time under ideal condition; ** physical travel distance coded by a network designer; A network designer determines Hour and Miles for dummy links.

In Scenario 2 (Table 4 and Figure 2b), the O-D pair between Prairie Rose and Moorhead resulted in two potential routes compared to four routes for the O-D pair between Prairie Rose and Fargo. Also note that ID 0 aligns well with the interstate highway system since the vehicle capacity over the interstate highway system is larger than for other routes. As such, ID 1 was assigned only a trip set size of 290 trips with a higher impedance value compared to that for ID 0. The O-D pair of Prairie Rose and Fargo encompasses six routes because local roads have less capacity than major roads even if route sections are parts of the interstate highway. The unassigned trips are reassigned to the second best route (ID 3) and to subsequent IDs. ID 5 shows an infeasible impedance value due to: (i) summed values designating congestion stemming from residuals from previously selected routes (ID 2 & ID 3); and (ii) connectivity based on an infeasible route since it is the only accessible route to the destination. ID 5 is a rare case since impedance is very high so it is infeasible even though the length is shorter than for other routes (ID 2 & 3).

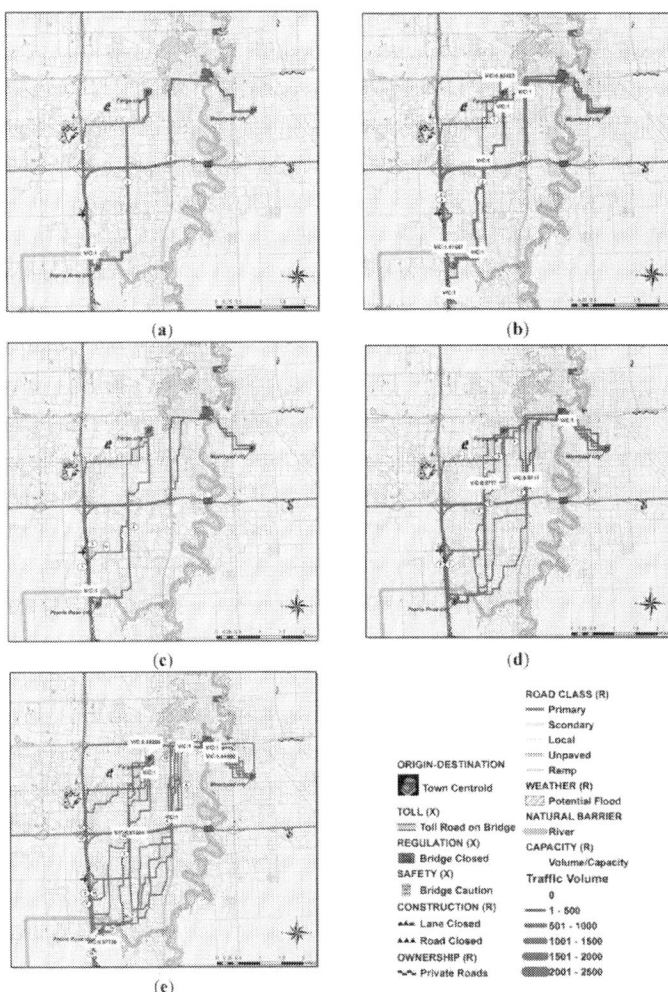

Figure 2. Selected routes from Scenarios 1–5. (**a**) Selected routes from Scenario 1; (**b**) Selected routes from Scenario 2; (**c**) Selected routes from Scenario 3; (**d**) Selected routes from Scenario 4; (**e**) Selected routes from Scenario 5.

Effects of Congestion

In Scenario 3, five routes are selected; two routes for the first O-D pairs and the other three routes for the second O-D pair (Table 4). Major highways would be chosen in general, but an alternative route was selected with low impedance and no congestion. Congestion impedance in this case may also function as a quasi-dynamic factor for route choice. Some segments double up as impedance multipliers due to (i) increased capacity; (ii) a higher road classification; and (iii) a lower relative impedance.

In Scenario 4, newer routes are selected for the O-D pairs between Prairie Rose and Moorhead devoid of capacity conflict even though impedance values are different (Table 4). The route ID 2 has higher impedance than ID 1; however, the total length in miles is shorter. This indicates that the route with ID 2 has a higher cost than ID 1 because hierarchically it is the route with the third lowest impedance. The Prairie Rose-Fargo routes also show that the shortest paths had higher impedances.

Effects of Packet Size

Nineteen routes were distributed across the region with primary roads selected repeatedly as indicated by overlapping (thicker) lines and assigned trip information (Table 4). This mimics stochastic user equilibrium model based on impedance information. The subdivided trips were sensitive to (a) congestion; and (b) stochastically-derived impedances values. From Figure 2a,b, dissimilar trends for travel length and impedance can be observed from the Prairie Rose and Moorhead pair. Figure 2c,d display similar incremental trends for travel length and impedance from the Prairie Rose and Fargo pair. Some points that may need to be addressed in the future simulations may include the following:

- Trip sequence and O-D information may have an impact on the route choice, which can further be constrained by capacity and other factors.
- A cost set is important for better simulation and design such as annual average daily traffic (AADT) data from highway performance management systems (HPMS) [5].

Discussion

In this study, we have broken down impedance into absolute and relative impedance and also into links and nodes in response to infrastructure and would-be users' preferences (e.g., volume/capacity ratio, hazards, bridge operation, construction, right-of-way, and road classification) [32]. Physical attributes, such as locations (nodes) and roads (links), and pertinent dynamics were deemed within the scope of an ideal routing system. We developed an impedance-based multi-criteria decision-making algorithm using GIS, thereby extending a previous single impedance approach for traffic assignment. Five scenarios were developed to evaluate and illustrate the algorithm under set criterion including FIFO assignment and capacity constraints.

We found that considering only distance or travel time as separate entities was not realistic. Other factors needed to be accounted for by adopting induced impedances. When distance is weighted appropriately using the possibility value to transect a segment, a modified distance value can then be recalculated to simulate realistic traffic impacts. From the Oak Ridge National Laboratory database [33], a dummy link was assigned to any terminal that is physically unconnected to a segment or a node in the contiguous layer. In our approach, a bridge exists as a node on a segment without linkage to the segment or node. So, to account for a bridge closure, a logical construct was assigned for the affected segment in order to generate a node impedance value. Gentile and Papola [34] used a similar segment level approach rather than an aggregate path because they felt that a path by itself is not an elementary unit for proper route choice and traffic assignment. In this regard, link impedance can best be applied to each segment. Accessibility is therefore narrowed to lower impedance segments to ease traversing from a selected origin to destination.

Three different types of traversability are employed: "go" for low impedance values of 1, "detour" for high impedance values of 0, and "slow" for impedance values between 0 and 1. For instance, if roads are not blocked or unregulated by policies, vehicles can pass through the selected routes as a "go". Particular situations, such as accidents, severe weather, congestion, and poor level of service (LOS), can result into a "slow" trip. The "slow" situation includes delay and results in an incurred penalty cost. The "detour" case occurs from absolute barriers such as the lack of capacity, bridge collapse, road blocking/closing, regulation, and congestion. Some relative impedance is closely related to "slow" and "detour" coincidentally to a certain degree like severe traffic accidents, congestion, and severe weather conditions.

A high level of flexibility and agility to respond to social and environmental effects is required in emerging markets and transportation sectors. Some impedance is inevitable (value added) and some undesirable (non-value added), causing higher total transportation costs and slower speeds. The approach in this study can also be used to gauge a disruption scenario. This can be done by altering the impedance value in a network and identifying critical segments and vulnerable infrastructure. This multi-attribute impedance traffic assignment model can be extended to freight shipments that may be affected by street closures such as was the case in the 2013 Boston Marathon. On that fateful day significant streets were blocked due to terrorism fears, significantly affecting scheduled shipments. This algorithm can also be used in conjunction with route user surveys to extend its functionality. The distance weighted impedance can easily be replaced with travel time weighted impedance in similar studies.

CONCLUSIONS

In this study, classical impedance was determined and then categorized into link impedance and node impedance. This was done to develop an innovative multi-attribute impedance-based modified all-or-nothing traffic assignment model. The advantage of this approach is to eliminate the efforts of removing unconnected segments physically and logically at the pre-processing stage of network analysis. In doing so, the multi-attribute impedance can be easily controlled and manipulated in a GIS. Thus, GIS tools can be utilized for travel demand modeling in a long-range transportation planning. To establish an optimum route, the multi-attribute impedance was minimized by decreasing either segment and/or node impedance. In addition, the method proposed in this study offers flexibility in applied transportation and logistics studies and provides a means for analysts to respond promptly to inherent variables. To evaluate a model network, complex dynamic models allied with weather information, congestion, seasonal regulation areas, and various geographical factors can be implemented using attributes of the proposed approach. In addition, the algorithm discussed in this study can be used to reflect decision-making rules and behavior such as risk-averse, risk-neutral, and risk-taker drivers by utilizing stochastically generated user behaviors. Thus, although the experiments were done using all-or-nothing, this multi-attribute impedance concept can be used with a typical path-based traffic assignment algorithm.

Some limitations of the study are found. Non-linear impedance values should be further tested for the route selection process. In addition, a dynamic stochastic trip assignment model with trajectory should be considered in the future to incorporate congestion and dynamic factors such as accident, and other unforeseeable events for the short-term operations planning. In the future, a linear traversability and impedance model should be calibrated in order to respond to travelers' behavior. In addition, the level of attributes and traversability can be determined by using fuzzy inference [32] with linguistic classes. Advanced survey instruments such as optical license plate recognition and global positing systems (GPS) tracking systems can be utilized to predict driver's behavior and combined with the multi decision factors for assigning trips to the roads networks.

ACKNOWLEDGMENTS

We thank the anonymous reviewers who reviewed this manuscript. The authors would like to acknowledge funding from Mountain-Plains Consortium, Regional Transportation Center, Transportation and Research and Innovative Technology Administration, U.S. Department of Transportation. The results and opinions in this paper are those of the authors and do not necessarily reflect the policy of the sponsors.

AUTHOR CONTRIBUTIONS

EunSu Lee conceived and designed the experiments and wrote the most part of the paper, Peter G. Oduor contributed analysis tool and partially wrote the paper and provided the framework of the analysis.

REFERENCES

1. Wardrop, J.G. Some theoretical aspects of road traffic research. ICE Proc. Eng. Divisions **1952**, 1, 325–362.

2. Daganzo, C.F.; Sheffi, Y. On stochastic models of traffic assignment. Transp. Sci.**1997**, 11, 253–274.

3. Damberg, O.; Lundgren, J.T.; Patriksson, M. An algorithm for the stochastic user equilibrium problem. Transp. Res. Part B **1996**, 30, 115–131.

4. Bell, M.G.H.; Shield, C. A stochastic user equilibrium path flow estimator. Transp. Res. Part C **1997**, 5, 197–210.

5. Battelle. FAF3 Freight Traffic Analysis. Available online: http://faf.ornl.gov/fafweb/Data/Freight_Traffic_Analysis/faf_fta.pdf (accessed on 18 May 2015).

6. Dijkstra, E.W. A note on two problems in connexion with graphs. Numer. Math.**1959**, 1, 269–271.

7. Tolliver, D.; Dybing, A.; Lu, P.; Lee, E. Modeling investments in county roads and local roads to support agricultural logistics. J. Transp. Res. Forum **2011**, 50, 101–115.

8. Lee, E.; Oduor, P.G.; Farahmand, K.; Tolliver, D. Heuristic path-enumeration approach for container trip generation and assignment. J. Transp. Res. Forum **2011**,12, 7–21.

9. Petersen, E.R. A primal-dual trip assignment algorithm. Manag. Sci. **1975**, 22, 87–95.

10. Janic, M. Modeling the full costs of an intermodal and road freight transport network.Transp. Res. Part D **2007**, 12, 33–44.

11. Keshkamat, S.S.; Looijen, J.M.; Zuidgeest, M.H.P. The formulation and evaluation of transport route planning alternative-s: A spatial decision support system for the Via Baltica Project, Poland. J. Transp. Geogr. **2009**, 17, 54–64.

12. Geurs, K.T.; Wee, B.B. Accessibility evaluation of land-use and transport strategies: Review and research directions. J. Transp. Geogr. **2004**, 12, 127–140.

13. Woudsma, C.; Jensen, J.F.; Kanaroglou, P.; Maoh, H. Logistics land use and the city: A spatial-temporal modeling approach. Transp. Res. Part E **2008**, 44, 277–297.

14. Shaw, S.L. What about "Time" in transportation geography? J. Transp. Geogr. **2006**,14, 237–240.

15. Miller, H.J. Measuring space-time accessibility benefits within transportation networks: Basic theory and computational procedures. Geogr. Anal. **1999**, 31, 187–212.

16. Weber, J.; Kwan, M. Bringing time back in: A study on the influence of travel time variations and facility opening hours on individual accessibility. Prof. Geogr. **2002**,54, 226–240.

17. Nesterov, Y.; de Palma, A.D. Stationary dynamic solutions in congested transportation networks: Summary and perspectives. Netw. Spat. Econ. **2003**, 3, 371–395.

18. Taylor, N.B. The CONTRAM dynamic trip assignment model. Netw. Spat. Econ.**2003**, 3, 297–322.
19. Lee, E.; Oduor, P.G.; Farahmand, K. Simplistic geospatial techniques in analyzing transportation dynamics for origin-destination container movement routes in the United States. J. Transp. Syst. Eng. Inf. Technol. **2012**, 12, 79–90.
20. Peeta, S.; Ziliaskopoulos, A.K. Foundations of dynamic trip assignment: The past, the present and the future. Netw. Spat. Econ. **2001**, 1, 233–265.
21. Carey, M. Nonconvexity of the dynamic trip assignment problem. Transp. Res. Part B **1992**, 26, 127–133.
22. Kwan, M.P.; Ransberger, D.M. LiDAR assisted emergency response: Detection of transport network obstructions caused by major disasters. Comput. Environ. Urban Syst. **2010**, 34, 179–188.
23. Grossardt, T.; Bailey, K.; Barumm, J. Analytic minimum impedance surface: Geographic information system-based corridor planning methodology. Transp. Res. Rec. **2001**.
24. Chen, T.Y.; Chang, H.L.; Tzeng, G.H. Using a weight-assessing model to identify route choice criteria and information effects. Transp. Res. Part A **2001**, 35, 197–224.
25. Cascetta, E. Transportation System Analysis: Models and Applications; Springer: New York, NY, USA, 2009.
26. Gutiérrez, J.; Urbano, P. Accessibility in the European Union: The impact of the trans-European road network. J. Transp. Geogr. **1996**, 4, 15–25.
27. Arnold, P.; Peeters, D.; Thomas, I. Modeling a rail/road intermodal transportation system. Transp. Res. Part E **1994**, 40, 255–270.
28. Handy, S.L. Regional versus Local Accessibility: Implications for Nonwork Travel. Available online: http://www.uctc.net/papers/234.pdf (accessed on 19 May 2015).
29. Transportation Research Board. Highway Capacity Manual; National Research Council: Washington, DC, USA, 2000.
30. Moseley, M.J. Accessibility: The Rural Challenge; Methuen and Company Limited: London, UK, 1979.
31. Stanilov, K. Accessibility and land use: The case of suburban Seattle, 1960–1990.Reg. Stud. **2003**, 37, 783–794.
32. Reddy, H.K.; Chakroborty, P. A Fuzzy inference based assignment algorithm to estimate O-D matrix from link volume counts. Comput. Environ. Urban Syst. **1998**,22, 409–423.
33. Oak Ridge National Laboratory. Railroad Network. Available online: http://cta.ornl.gov/transnet/RailRoads.html (accessed on 18 May 2015).
34. Gentile, G.; Papola, A. An alternative approach to route choice simulation: The sequential models. In Proceeding of the European Transport Conference, Strasbourg, France, 18–20 September 2006.

CITAION

EunSu Lee and Peter G. Oduor, Using Multi-Attribute Decision Factors for a Modified All-or-Nothing Traffic Assignment, doi:10.3390/ijgi4020883

CHAPTER 9

Characterization of Black Spot Zones For Vulnerable Road Users In São Paulo (Brazil) And Rome (Italy)

Cláudia A. Soares Machado [1,], Mariana Abrantes Giannotti [1], Francisco Chiaravalloti Neto [2], Antonino Tripodi [3], Luca Persia [3] and José Alberto Quintanilha [1]*

[1]Laboratory of Geoprocessing, Department of Transportation Engineering, Polytechnic School of the University of São Paulo, Av. Prof. Almeida Prado, Travessa 2, 83, SP 05508-070 São Paulo, Brazil; EMails: mariana.giannotti@gmail.com (M.A.G.); jaquinta@usp.br (J.A.Q.)

[2]Department of Epidemiology, School of Public Health of the University of São Paulo, Av. Dr. Arnaldo, 715, SP 01246-904 São Paulo, Brazil; E-Mail: franciscochiara@usp.br

[3]Centro di Ricerca per il Trasporto e la Logistica, Università di Roma, Piazzale Aldo Moro, 5, Roma 00185, Italy; E-Mails: tripodi@ctl.uniroma1.it (A.T.); luca.persia@uniroma1.it (L.P.)

ABSTRACT

Non-motorized transportation modes, especially cycling and walking, offer numerous benefits, including improvements in the livability of cities, healthy physical activity, efficient urban transportation systems, less traffic congestion, less noise pollution, clean air, less impact on climate change and decreases in the incidence of diseases related to vehicular emissions. Considering the substantial number of short-distance trips, the time consumed in traffic jams, the higher costs for parking vehicles and restrictions in central business districts, many commuters have found that non-motorized modes of transportation serve as viable and economical transport alternatives. Thus, local governments should encourage and stimulate non-motorized modes of transportation. In return, governments must provide safe conditions for these forms of transportation, and motorized vehicle users must respect and coexist with pedestrians and cyclists, which are the most vulnerable users of the transportation system. Although current trends in sustainable transport aim to encourage and stimulate non-motorized modes of transportation that are socially more efficient than motorized transportation, few to no safety policies have been implemented regarding vulnerable road users (VRU), mainly in large urban centers. Due to the spatial nature of the data used in transport-related studies, geospatial technologies provide a powerful analytical method for studying VRU safety frameworks through the use of spatial analysis. In this article, spatial analysis is used to determine the locations of regions that are characterized by a concentration of traffic accidents (black zones) involving VRU (injuries and casualties) in São Paulo, Brazil (developing country), and Rome, Italy (developed country). The black zones are investigated to obtain spatial patterns that can cause multiple accidents. A method based on kernel density estimation (KDE) is used to compare the two cities and show economic, social, cultural, demographic and geographic differences and/or similarities and how these factors are linked to the locations of VRU traffic accidents. Multivariate regression analyses (ordinary least squares (OLS) models and spatial regression models) are performed to investigate spatial correlations, to understand the dynamics of VRU road accidents in São Paulo and Rome and to detect factors (variables) that contribute to the occurrences of these events, such as the presence of trip generator hubs (TGH), the number of generated urban trips and demographic data. The adopted methodology presents satisfactory results for identifying and delimiting black spots and establishing a link between VRU traffic accident rates and TGH (hospitals, universities and retail shopping centers) and demographic and transport-related data.

INTRODUCTION

One key element of modern transportation systems is safety. The goal of safety is to minimize the number of accidents and to reduce the severity of injuries for all users, including motorists, passengers of particular vehicles, public transport commuters, cyclist and pedestrians [1,2].

Traffic accidents result in the second highest cost of transportation. These costs result from personal damage (injuries and wounds), fatalities, property damage (to vehicles and other public or private property), degradation of quality of life and decreases in available time for conducting activities and maintaining social relationships [3,4,5].

Non-motorized modes of transportation, especially cycling and walking, offer numerous benefits, including improvements in the livability of cities, healthy physical activity, an efficient urban transportation system, less traffic congestion, less noise pollution, clean air, less impact on climate change, decreased incidence of diseases related to vehicular emissions, decreased fossil fuel use and decreased transportation costs. Considering the substantial number of short-distance trips, the time consumed in traffic jams, higher vehicle parking costs and restrictions in central business districts, many commuters have found that non-motorized modes of transportation are viable and economical transportation alternatives [6].

Thus, non-motorized modes of transportation should be stimulated and encouraged by local governments. In return, governments must provide safe conditions for these users and the users of motorized transport that allow motorized transportation to exist in harmony with pedestrians and cyclists. Planners and engineers must accommodate the needs of cyclists and pedestrians by designing transportation facilities for urban areas [7].

It is widely recognized that pedestrians and cyclists are the most vulnerable users of the transportation system. Although current trends in sustainable transport aim to encourage and stimulate non-motorized modes of transportation that are socially more efficient than motorized modes, few to no safety policies related to vulnerable road users (VRU) (traffic participants without outer protective cells) [8] have been implemented in large urban centers. Therefore, it is important to promote a conceptual road safety framework to reduce and control accident risks involving VRU. The risks (crash risk and injury severity) result from travel behavior (volume, modal split and distribution of traffic over time and space) and the characteristics of the transport infrastructure, such as the type of vehicle and road user [9].

The road safety framework for VRU is affected by land use and urban infrastructure (built environment). This framework is an abstraction or simplification of reality that can be used to help planners, policy makers and decision makers better understand real-world systems, facilitate communication and integrate knowledge across a variety of engineering and scientific disciplines, including transportation science, geography, urban planning, economics and physics [9,10].

Due to the spatial nature of the data involved in transport-related studies, geospatial technologies provide a powerful analytical method for studying VRU safety frameworks through the use of spatial analysis. The rapid development of geographic information science and its related technologies has resulted in the collection of ample transportation data to better understand road traffic accident patterns and the behaviors of transportation system users [10]. In addition, the availability of real-time traffic data obtained through geospatial technologies increased and proactively stimulated proactive safety management in transport networks [11]. In contrast with conventional approaches, geospatial methods are used to analyze the spatial patterns of accident locations within a network space and are not affected by the configuration of the street network or its distance [12]. During the last few years, the identification of hazardous locations, which are called black spot zones, in road networks has substantially progressed. The identification of black spot zones is facilitated by the application of geospatial technologies in transportation research, which has enabled precise localization of traffic accidents and the identification of spatial patterns involving regions with a high occurrence of traffic accidents [13].

The objective of this study is to determine the locations of the regions (that contain the critical road sections) that are characterized by a concentration of traffic accidents (black zones) involving VRU (injuries and casualties) in São Paulo, Brazil (emerging country), and Rome, Italy (developed country). The black zones are investigated to obtain spatial patterns that result in multiple accidents. A method based on kernel density estimation (KDE) is applied to analyze economic, social, cultural, demographic and geographic patterns of urban road accidents in São Paulo and Rome and to investigate how these factors are linked to the locations of VRU traffic accidents. Moreover, to understand the spatial interactions that occur between VRU traffic accidents and their locations, multivariate regression analyses (Ordinary Least Squares (OLS) models and spatial regression models) are performed. In addition, these analyses are performed to determine the most important attributes that contribute to the high rates of VRU traffic accidents in a given region.

Background

This literature review mainly discusses three aspects related to the spatial characterization of hot spots involving VRU, traffic accident research, accidents involving VRU and the spatial analysis of these accidents.

Traffic Accident Research

One of the most troubling problems of transportation systems is related to accidents. Road traffic accidents result in serious societal problems with significant individual, property and society costs [2,14]. The World Health

Organization (WHO) indicates that road traffic injuries comprise a major, but neglected global public health problem that requires action for effective and sustainable prevention [15]. According to the WHO, 1.24 million people were killed worldwide in 2010 due to traffic accidents. Middle-income countries, which are becoming motorized rapidly, are the hardest hit by traffic accidents, with approximately 70% of traffic-related deaths occurring in these countries [16,17]. This tragic scenario indicates that traffic accidents are a serious public health and welfare concern and can be considered as a global epidemic [12]. The cost of dealing with the consequences of these traffic accidents reaches billions of dollars [17], which is a large sum that could be used in the transportation system to prevent traffic accidents. Half of all road traffic deaths involve VRU [17]. In Brazil, 4% of the individuals that died from traffic accidents were cyclists and 23% were pedestrians in 2010 [18].

Land use environments influence the needs and behaviors of VRU. The choice of a transport mode (non-motorized or motorized) varies with the land use type. For example, the number of parked bicycles at transit stations, the percentages of land for commercial use, the distances between origins and destinations and the nearest bus stop with services serving the transit station affect transport mode decisions. These factors are influential when promoting non-motorized transport. Moreover, the most significant concern regarding non-motorized transportation is the risk of traffic accidents. The accident risk varies with the level and type of local traffic. Traffic accident risk is the most important concern for VRU in urban centers [19].

In recent years, an increasing number of research studies have been conducted regarding traffic accident patterns through spatial approaches. An approach (similar to that developed in this study) was conducted by [20] that identified and delimited road sections that were characterized by concentrations of traffic accidents (black zones) by applying and comparing two methods, local spatial autocorrelation indices (a decomposition of the Global Moran Index) and KDE, without any reference to the spatial patterns of the accidents. In the study presented by [21] was used a Geographic Information System (GIS) and log-linear model to investigate the spatial distribution of pedestrian/cyclist accidents involving school-aged children in Florida, USA, and examined the conditions under which these events were more likely to occur. In the research performed by [13] was developed a procedure for identifying and evaluating clusters (black spot zones) of traffic accidents by using the KDE in the Southern Moravia Region of the Czech Republic. A spatial and temporal analysis of VRU traffic accidents in Santiago, Chile, was conducted by [22], in order to identify the most critical areas (black spot zones) in a GIS environment (using the KDE and Moran Index), and also to detect the attributes and the contributing factors (for example, time of day, straight road sections and intersections and roads without traffic signs within the critical areas) associated with VRU traffic accidents. The relationships between three years of pedestrian crash counts across census tracts in Austin, Texas (United States), and several land use networks and demographic attributes, such as land use balance, residents' access to commercial land uses, sidewalk density, lane-mile densities (by roadway class) and population and employment densities (by type) were examined by [23]. A study was presented by [24], which integrated a spatial density analysis (KDE) and the local Moran Index to detect black spot zones of traffic accidents and to formally evaluate the extensiveness of locations with high densities to provide tools and information for planners and decision makers to effectively allocate resources for accident prevention and safety improvement. A spatial-temporal analysis of road accidents in New Brunswick, Canada, to study the impacts of climate change on hazardous weather-related traffic accidents was provided by [25]. In addition, a spatial analysis (KDE and wavelet analysis) to identify black spot regions of traffic accidents and to verify patterns that contributed to accidents that occurred in the black spot zones in Paraná State, Brazil, was applied by [26]. A spatiotemporal analysis of intra-urban traffic accidents in metropolitan Shiraz, Iran, was presented by [27], whose objective was to identify accident-prone zones and sensitive hours using GIS-based spatio-temporal visualization techniques. This analysis is aimed at identifying high-rate accident locations and safety deficient areas by using the KED method. A spatial Bayesian modeling approach was proposed by [28] to predict VRU accident risks for a road network and to identify how road infrastructures influence VRU safety in Brussels, Belgium. An approach was introduced by [29] for the identification of hazardous accident zones that compares spatial and non-spatial methods. Overall, the study concludes that spatial analysis methods outperform non-spatial approaches, because they do not require the segmentation of highways. The only information that is required when using the spatial analysis method to identify black spot zones is the location of each accident.

VRU Traffic Accidents

In developing countries, accidents involving VRU constitute a much larger fraction of all traffic-related fatalities and injuries than those in developed countries. Studies and observations have indicated that a large fraction of VRU injuries and fatalities occur at urban intersections (approximately one-third) and in collisions with automobiles (about two-thirds). Most VRU accidents are caused by a combination of behavioral, technological and environmental factors, which indicates that the safety of VRU should be improved. In developing countries, safety can be improved by adopting safety principles that have reduced the number of injuries and casualties in VRU traffic accidents in developed countries, such as adequate road design and traffic management [30,31].

The characteristics of mixed traffic, the inadequate driving abilities of vehicle drivers, the behavior of users of the transportation system (the drivers of motorized and non-motorized vehicles, passengers and pedestrians), vehicle factors (such as insufficient maintenance), poor design and layout of roads and other transport infrastructure and weather conditions frequently render traffic in a country unsafe and can influence the occurrence of accidents. While these factors are frequent in low-/middle-income countries, because high-quality automotive engineering, road design and vehicles are continuously being optimized in developed countries in terms of safety, human causes are considered as the main causes of VRU traffic accidents. Thus, the number and incidence of traffic accidents involving VRU have different economic consequences in developing countries (low- and middle-income countries) and developed countries (high-income countries) [8].

Public policies (government budget and engagement for transport safety and the level of enforcement of road traffic rules), personal characteristics (health and physical condition, income, education, age and gender), socio-economic conditions (economic growth) and demographic and geographic circumstances (average distance traveled, the amount of time traveled and the population density) are factors that influence the rates of VRU traffic accidents [16]. Consequently, understanding these various factors and identifying their separate and/or combined effects on accident frequency and severity is important [11].

Spatial Analysis of VRU Accidents

To identify hotspots of road accidents (black spot zones) is important for determining effective strategies for reducing areas with a high density of accidents and for appropriately allocating resources for improving safety [32]. The characteristics of spatial analysis (GIS environment) for managing and processing locational and related information provide a robust understanding of the indicators of casual effects. Among several techniques, the KDE technique is one of the most frequently-used spatial tools for studying traffic accident phenomena, as demonstrated by [2,13,24,27,32,33,34,35,36,37,38,39,40].

The KDE technique is used to calculate the probability density function of a distribution from which a sample has been observed by centering a probability density function on each of the observed events [41]. The kernel estimator is a non-parametric algorithm that uses a density estimation method. This technique allows one to evaluate the local probability accident occurrence and, consequently, the probable dangerousness of a spatial unit [20]. KDE is the most widely-used nonparametric method in recent decades [42]. The KDE describes the distribution of the location of an event and ignores its association with values. This distribution is characterized by the density of events that occur around a centroid and represents the behavioral patterns of points or lines. In this study, the events are the locations (geocoded) of the accidents (represented by points), and the KDE is used to calculate the probability density function of each accident location.

Kernel density analysis is performed by passing a moving window over the data, usually on a regular grid. The densities of the observations within a set radius are calculated for each event located on the grid, and the contributions of each observation are weighted by its proximity to the center of the moving window. Thus, the result of applying KDE is a density map (raster format). The values of each pixel represent the relationships between the concentrations of the events per unit area. In addition, KDE can be used to calculate the density of punctual events (i.e., the density of traffic accidents in a region) or linear events (i.e., the density of a road network in a zone). It is important to highlight the simplicity, satisfactory properties and good results of the KDE method [43,44].

The areas where events are concentrated are identified by KDE analysis and have the highest accident rates involving VRU. These areas are called black spots (zones that reveal concentrations of accidents). The existence of black spot zones results from the awareness of the spatial interactions between contiguous traffic accident locations. The most straightforward use of GIS for accident analysis is the examination of spatial characteristics and attributes of traffic accident locations. In fact, the use of GIS has several advantages of the use of non-spatial methods for accident analysis [45].

To determine the most significant variables in urban systems (sociodemographic and transport-related factors) that are involved in the occurrence of road accidents, several researchers have used multivariate regression analyses, including [46,47,48,49,50,51,52,53,54,55,56].

In this study, the quantity of VRU traffic accidents was regressed against the number of explanatory variables by using two models, the ordinary least squares (OLS) model and the spatial autoregressive lag (SAR) model.

The traditional OLS multiple regression model is relatively quick, simple and suitable for analyzing punctual events, such as road accidents. As a global regression method, an important assumption of the OLS model is that all variables are stationary across the study area. In addition, the OLS method can involve potential issues related to spatial and temporal autocorrelations, and the endogeneity between the dependent variable (amount of road accidents) and independent variables (such as, population, area, income, trips generated, etc.) is possible [57,58,59]. In turn, SAR is a strong modeling approach that can be adopted when a spatial autocorrelation is highly likely. In the SAR method, the interactions are modeled as a weighted average of the neighboring observations. The endogenous variables that comprise the interactions are called spatially-lagged dependent variables, and the weights, which are grouped into a neighborhood matrix (by contiguity or k nearest neighbors), form the distinctive core of the class of spatial process models. These two methods (OLS and SAR) provide comparisons across different specifications, tests for robustness and allow one to capture the importance of spatial interactions and interdependencies among the involved attributes [60,61].

STUDY AREAS

One study area is São Paulo City (Figure 1), which is the capital of São Paulo State and is located in Southeastern Brazil. São Paulo is an excellent example of a rapidly-growing city, with more than 11.3 million people (2013) in an area of 1521.101 km^2, a demographic density of 7398.26 inhabitants/km^2, an urbanization rate of 99.10% (2010) and a geometric growth rate of 0.59% (during the period 2010–2013). According to the United Nations Organization (UNO), São Paulo is one of the 27 megacities in the world (source: IBGE—Instituto Brasileiro de Geografia e Estatística).

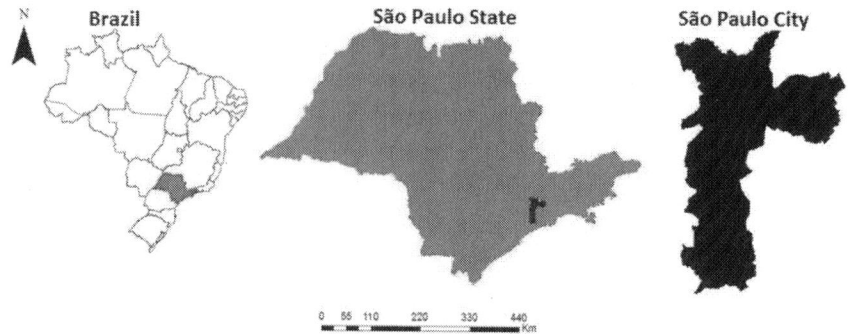

Figure 1. Study area: São Paulo. Source: [2].

The second study area, Rome (Figure 2), is a city and special commune (named "Roma Capitale") in Italy. Rome is the capital of Italy, the Province of Rome and the region of Lazio. Rome is Italy's largest and most populated commune and the fourth most populous city in the European Union. Its population (2013) is 2,913,349 inhabitants, with an area of 1285.31 km^2 and a demographic density of 2266.7 inhabitants/km^2 (source: Roma Capitale, Annuario Statistico 2013, Unità Operative (U.O.) Statistica, Sistema Statistico Nazionale).

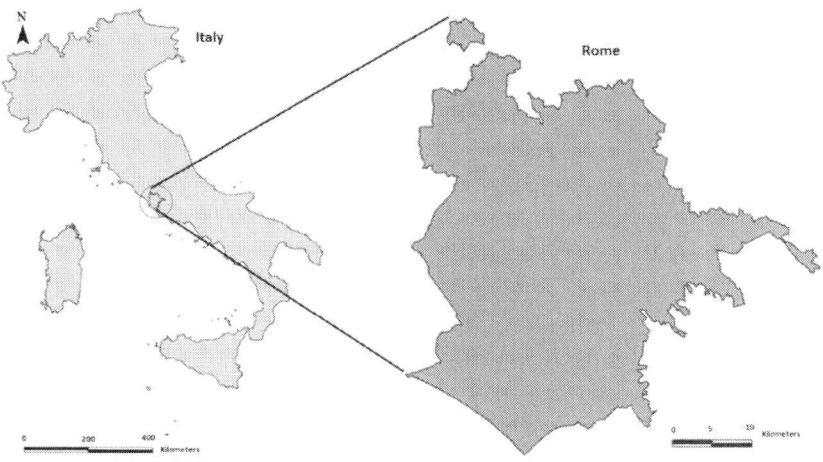

Figure 2. Study area: Rome. Source: Adapted from [62].

DESCRIPTION OF DATA AND METHODS

This study used a set of spatial and non-spatial data from São Paulo and Rome, which is described below.

São Paulo

The Municipality of São Paulo delivered the spatial unit adopted in the analysis (districts; Figure 3a) and the vectorial data (geocoded) that provided the locations of hospitals, universities and colleges, parks and recreational areas and retail shopping centers. The CET-SP (Companhia de Engenharia de Tráfego de São Paulo: the operative body of the municipality of São Paulo for traffic management of the city) provided the locations of traffic accidents involving VRUs in 2012 and in São Paulo's road network. Finally, the Metrô-SP (Companhia do Metropolitano do Estado de São Paulo: the public company responsible for operating the subway train system) performed the Origin and Destination Survey (2007 and 2012) that provided socioeconomic information and trip patterns.

Rome

The Research Center for Transport and Logistics (CLT) at the University of Rome delivered data regarding traffic accidents involving VRUs in 2012 with Rome's road network, socioeconomic information and vectorial data (geocoded), which provide the locations of hospitals, universities and colleges, parks and recreational areas and retail shopping centers. The spatial units adopted in this study are municipi, quartieri, rioni, suburbi and Zone Agro Romano (Figure 3b).

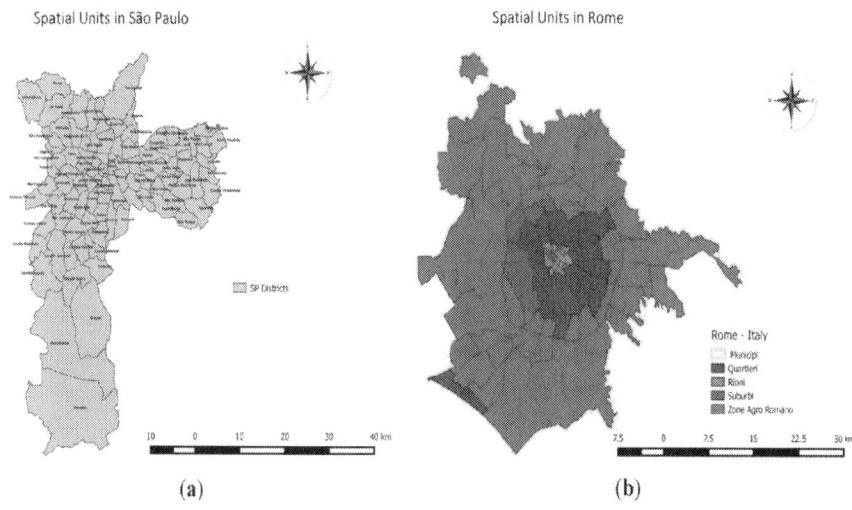

(a)   (b)

Figure 3. Spatial units: (**a**) São Paulo; and (**b**) Rome.

Methodology

The first step was to standardize the spatial data in terms of the coordinates systems, spatial units and spatial adjustment. Next, an analysis of VRU traffic accidents (total amount of accidents and the amount of fatal accidents, São Paulo and Rome; see Figure 4a,b) was performed in the GIS environment by using density analysis (KDE).

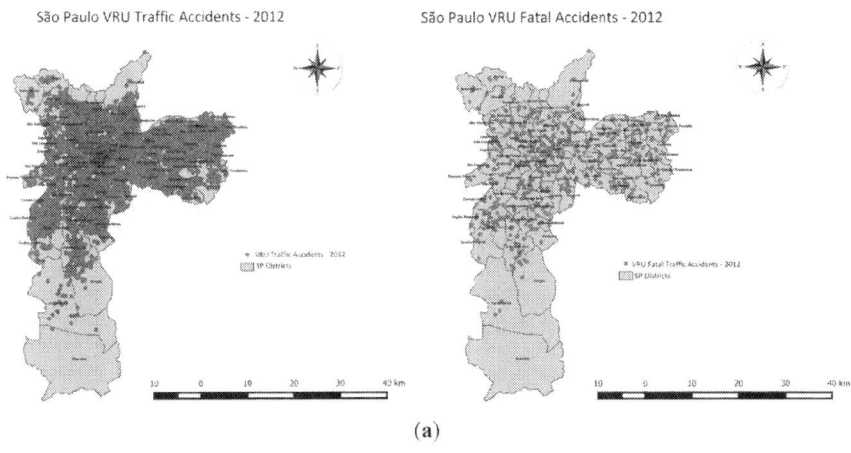

São Paulo VRU Traffic Accidents - 2012   São Paulo VRU Fatal Accidents - 2012

(a)

Figure 4. (a) Vulnerable road user (VRU) traffic accidents in São Paulo; (b) VRU traffic accidents in Rome.

The objectives of KDE analysis were to identify: (1) the spatial locations of black spots; (2) the attributes of each accident inside the black spot zones (address, road type, accident type, vehicle(s) involved and date/time); (3) the attributes of the region delimited by the black spots (area, population, demographic density and road network density); and (4) the number and spatial locations of the entities considered in this study as trip generator hubs (TGH) (hospitals, universities and colleges and retail shopping centers) inside the black spot zones.

An ecological study was conduct using data aggregated by spatial units (in São Paulo, districts; in Rome, quartieri, rioni, suburbi and Zone Agro-Romano; see Figure 3) to investigate the spatial clusters (black spot zones) and possible associations between the occurrences of urban road accidents and socio-economic and transport-related factors.

In the statistical analysis (spatial regression models), the following variables (in each spatial unit) were considered: the number of total and fatal VRU accidents; area; population; demographic density; the number of total/fatal accidents divided by area (VRU traffic accidents/km²); the number of total/fatal accidents divided by the population (VRU traffic accident accidents/inhabitants); the number of generated trips; the number of generated trips divided by area (trips/km²); the number of generated trips divided by population (trips/inhabitant); average income; per capita income; the number of retail shopping centers, hospital and universities and colleges; and a variable named TGH (trip generator hubs), which is calculated as the sum of the number of retail shopping centers, hospitals and universities and colleges. The first step was to calculate the correlation matrix (Spearman correlations) to identify collinear variables and to remove them from the regression model (correlation greater than 60%).

A regression analysis (OLS method) of the amount of total/fatal accidents divided by area (dependent variables) with the selected independent variables was performed. The variables incorporated in the model were selected by assessing the significance level (p-value). Possible explanatory variables were defined as those that presented p-values of ≤0.20 [63]. A global spatial analysis was also performed to verify the spatial dependence and variability around the predicted value (i.e., the spatial distribution and heterogeneity in model residuals) [64]. The importance of normal distributions is undeniable when applying regression models, because interpretation and inferences may not be reliable or valid when the normality assumption is violated [65]. Next, the Kolmogorov–Smirnov test for normality was performed, as described by [66].

Global Moran's Index is the most commonly-used test for global spatial autocorrelation [67], was determined for regression residuals (for more details, see [68]) to investigate the impacts of neighborhood matrix type and was used as a queen type of regular contiguity matrix (for more details, see [69]. When the Global Moran's Index does not present a significant positive spatial autocorrelation, the OLS model is considered appropriate. Otherwise, a two-step Lagrange Multiplier Test (LMT) is applied to reveal spatial dependence and spatial heterogeneity. In this study, when the Moran's Index indicates that the residuals present spatial dependence, the SAR model (for more details, see [70]) was used.

RESULTS

Kernel Density Estimator

The first analysis was the KDE, which was used in a GIS environment to identify black spot zones of traffic accidents (the total number of accidents and fatal accidents) involving pedestrians and cyclists. Due to the high number of road accidents in São Paulo (as shown in Figure 4a), the KDE generated four representative density classes for total accidents, low-, medium-, high- and very high-density accident areas (Figure 5). In addition, three density classes were generated for fatal accidents (Figure

6). In Rome, the quantity of road accidents was much lower than in São Paulo (Figure 4b). Thus, the KDE generated three density classes for total and fatal accidents, low-, medium- and high-density accident areas (Figure 7 and Figure 8).

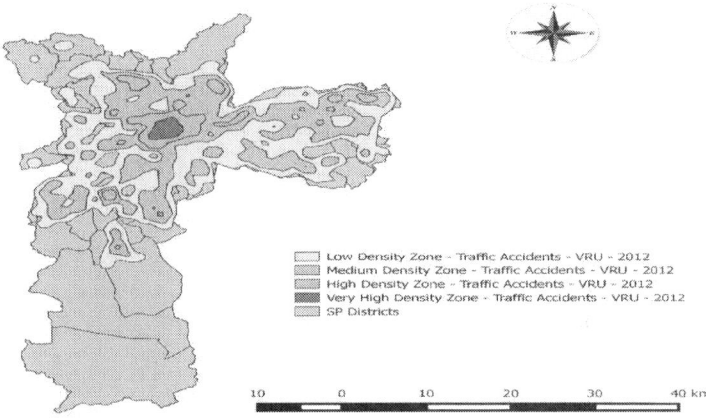

São Paulo VRU Traffic Accidents - 2012 - Black Spot Zones

Figure 5. Black spot zones. Traffic accidents involving VRU in São Paulo in 2012.

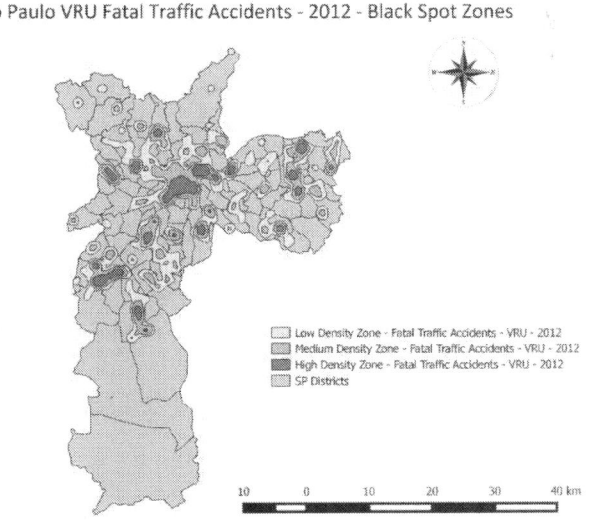

São Paulo VRU Fatal Traffic Accidents - 2012 - Black Spot Zones

Figure 6. Black spot zones. Fatal traffic accidents involving VRU in São Paulo in 2012.

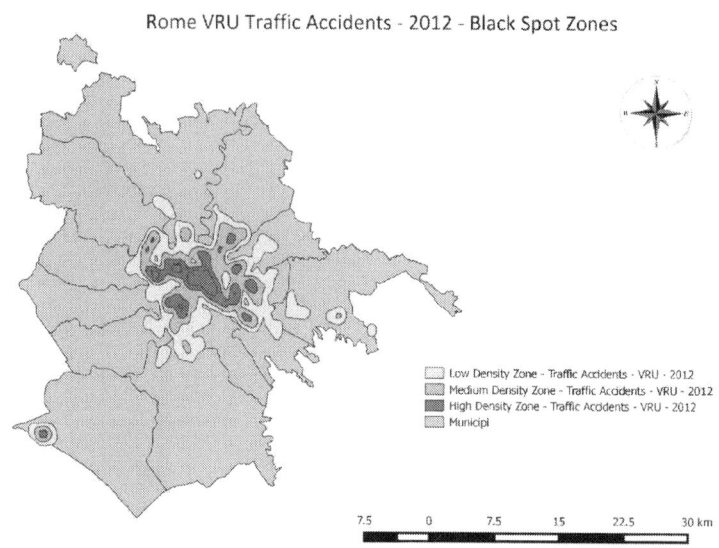

Figure 7. Black spot zones. Traffic accidents involving VRU in Rome in 2012.

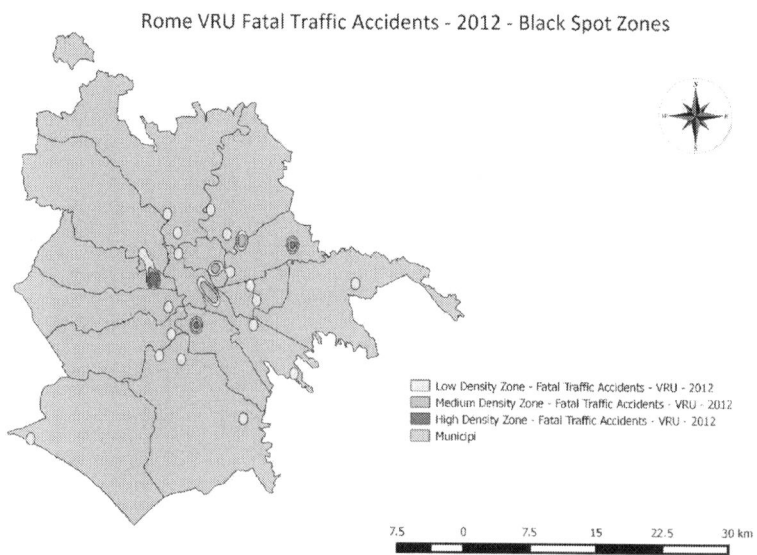

Figure 8. Black spot zones. Fatal traffic accidents involving VRU in Rome in 2012.

Statistical Analysis

The black spot zones were linked to the presence of TGH ((1) hospitals; (2) universities and colleges; and (3) retail shopping centers). Figure 9, Figure 10 and Figure 11 show the presence of TGH inside the density zones in São Paulo, and Figure 12, Figure 13 and Figure 14 show those of Rome.

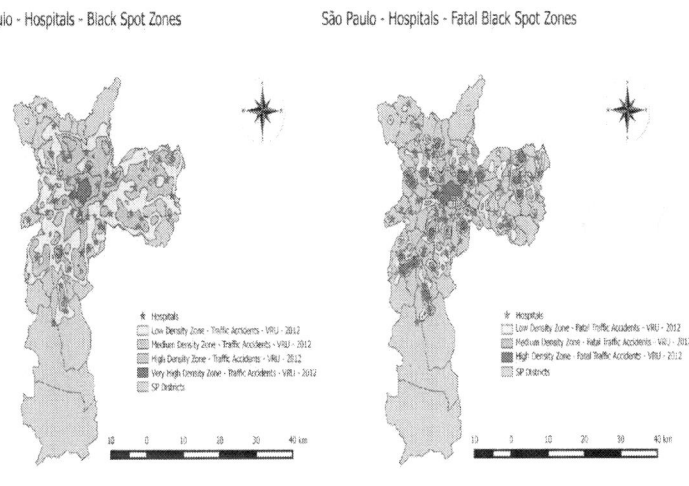

Figure 9. Hospitals/black spot: São Paulo.

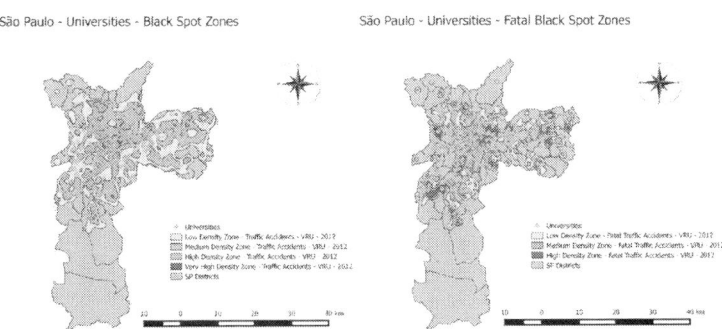

Figure 10. Universities /black spots: São Paulo.

Figure 11. Retail shopping centers/black spots: São Paulo.

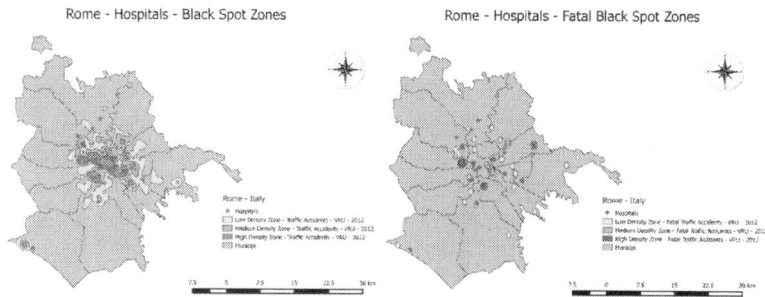

Figure 12. Hospitals/black spot: Rome.

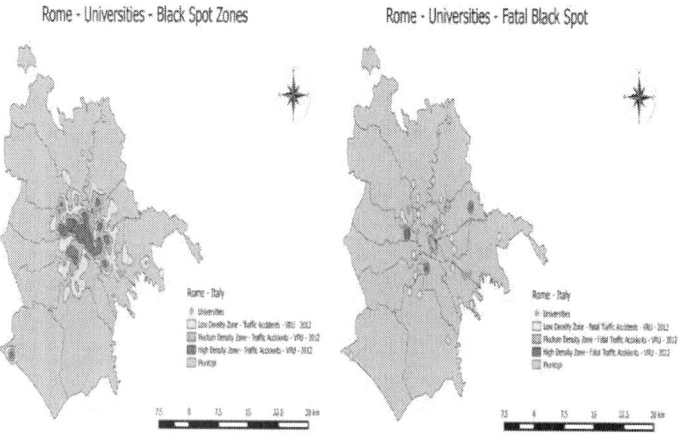

Figure 13. Universities and colleges/black spots: Rome.

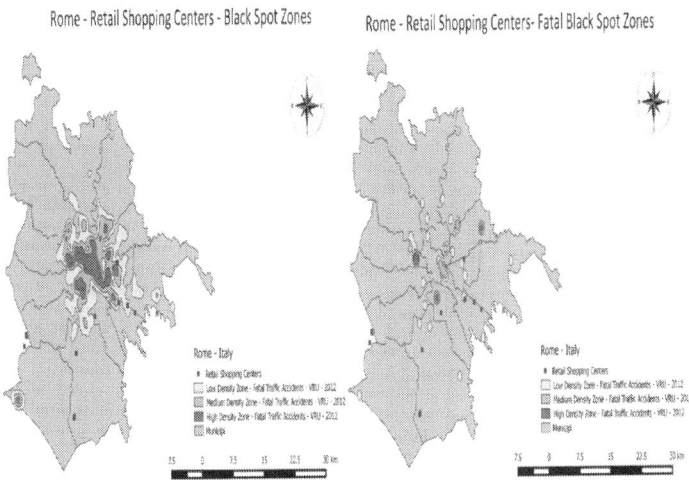

Figure 14. Retail shopping centers/black spots: Rome.

A visual analysis alone is not sufficient for affirming the existence of direct influences of TGH over black spot zones. Thus, we investigated whether the concentrations of these entities (TGH) are related by using high rates of traffic accidents in both cities and determined the variables that could be associated with the occurrences of these events.

São Paulo: Total VRU Traffic Accidents

In São Paulo, the explanatory variables that best explained the concentrations of total VRU traffic accidents (represented in the model by the rate number of total traffic accidents divided by area) included the number of generated trips divided by population (trips/inhabitant) and THG (retail shopping centers + hospitals + universities/colleges). The Kolmogorov-Smirnov test was used to assess the normality, and the fourth root of the total VRU traffic accidents was used to improve the normal distribution approximation. The results are presented below in Table 1 and show that the OLS model is appropriate.

Table 1. OLS regression model: São Paulo, total VRU traffic accidents.

Explanatory Variable	Regression Coefficient	Standard Error	t-Value	p-Value
Intercept	1.3852	0.0618	22.434	0.0000
Trips/inhabitants	0.0352	0.0098	3.575	0.0006
TGH (Trip Generator Hub)	0.0140	0.0045	3.084	0.0027

Residual standard error: 0.3611

Multiple R-squared: 0.2906

Adjusted R-squared: 0.2754

F-statistic: 19.05

p-value: 0.0000

Residual analysis:

Observed Moran's Index: −0.0221

$p = 0.8609$ (Moran test)

Expectation: −0.0106

Variance: 0.0043

Kolmogorov–Smirnov test: $p = 0.1651$

São Paulo: Fatal VRU Traffic Accidents

In São Paulo, the explanatory variable that best explained the concentration of fatal VRU traffic accidents (represented in the model by the rate number of fatal traffic accidents divided by area) was calculated as the generated trips divided by the population (trips/inhabitant). The Kolmogorov-Smirnov test was used to assess normality, and the square root of fatal VRU traffic accidents was used to improve the approximation of the normal distribution. The results are presented below in Table 2 and show that the OLS model is appropriate.

Table 2. OLS regression model: São Paulo, fatal VRU traffic accidents.

Explanatory Variable	Regression Coefficient	Standard Error	t-Value	p-Value
Intercept	0.4134	0.0513	8.051	0.0000
Trips/inhabitants	0.0426	0.0076	5.601	0.0000
Residual standard error: 0.3078				
Multiple R-squared: 0.2502				
Adjusted R-squared: 0.2423				
F-statistic: 31.37				
p-value: 0.0000				
Residual analysis:				
Observed Moran's Index: 0.1051				
p = 0.0664 (Moran test)				
Expectation: −0.0138				
Variance: 0.0042				
Kolmogorov-Smirnov test: p = 0.6124				

Rome: Total VRU Traffic Accidents

In Rome, the explanatory variable that best explains the concentration of total VRU traffic accidents is demographic density (inhabitants/km^2). The Kolmogorov-Smirnov test was used to assess the normality, and the square root of the total VRU traffic accidents was used to improve the approximation of the normal distribution. The results are presented below in the Table 3 and show that the OLS model is not appropriate. Then, the next step was the application of the SAR model.

Since Moran's Index is a significant value, the residuals present spatial dependence, and the OLS is not appropriate for modeling Rome's total VRU traffic accidents. Thus, the Lagrange Multiplier Test (LMT) was performed to select the best model. The results are shown in Table 4. When analyzing the LMT results, the Robust Lagrange Multiplier Spatially-Lagged (RLMlag) model was adopted.

The spatial autocorrelation lag model (RLMlag) results are presented in Table 5 and show that the SAR model is appropriate, because no residual spatial dependence occurs. The Akaike Information Criterion (AIC) (details in [71,72,73]) of the SAR model is lower than the AIC of the linear regression (OLS), which demonstrates that the SAR model is better than the OLS model for total VRU traffic accidents in Rome.

Table 3. OLS regression model: Rome, total VRU traffic accidents.

Explanatory Variable	Regression Coefficient	Standard Error	*t*-Value	*p*-Value
Intercept	0.6570	0.1627	4.038	0.0000
Demographic Density	0.0002	0.000002	10.589	0.0000
	Residual standard error: 1.317			
	Multiple R-squared: 0.4959			
	Adjusted R-squared: 0.4915			
	F-statistic: 112.1			
	p-value: 0.0000			
	Residual analysis:			
	Observed Moran's Index: 0.3758			
	p = 0.0000 (Moran test)			
	Expectation: −0.0113			
	Variance: 0.0022			
	Kolmogorov-Smirnov test: *p* = 0.7016			

Table 4. Lagrange Multiplier Test (LMT): Rome, total VRU traffic accidents.

Test	Statistic Value	*p*-Value
Lagrange Multiplier Error—LMerr	0.1058	0.0000
Lagrange Multiplier Spatially-Lagged—LMlag	3.1272	0.0000
Robust Lagrange Multiplier Error—RLMerr	5.9186	0.2344
Robust Lagrange Multiplier Spatially-Lagged—RLMlag	8.9401	0.0000
Spatial Autoregressive Moving Average—SARMA	9.0458	0.0109

Table 5. Spatial autocorrelation (SAR) lag model: Rome, total VRU traffic accidents.*Rome: Fatal VRU Traffic Accidents*

Explanatory Variable	Regression Coefficient	Standard Error	*z*-Value	*p*-Value
Intercept	−0.1444	0.1501	0.9616	0.3363
Demographic Density	0.0001	0.0000	7.9150	0.0000
	Rho: 0.63342			
	Likelihood Ratio Test - LR test value: 68.206			
	p-value: 0.0000			
	Asymptotic standard error: 0.068453			
	z-value: 9.2533			
	Wald statistic: 85.624			
	Log likelihood: −161.426 for lag model			
	ML residual variance (sigma squared): 0.88966 (sigma: 0.94322)			
	AIC: 330.85 (AIC for linear regression: 397.06)			
	LM test for residual autocorrelation			
	test value: 7.5693			
	p-value: 0.005937			

Due to the low occurrence of fatal traffic accidents involving VRU in Rome (only 34 events), it was not possible to perform a statistical analysis by using standard spatial regression models. In this case, it is necessary to adopt a zero-inflated model (a situation of excess zeros relative to what the standard models allow), because a large number of observations are equal to zero, as described in [74]. According to [75], zero-inflated count data refer to data for which a generalized linear model has a lack of fit due to too many zeros. Such data are common in many applications (in this study, the research involving urban road accidents), especially when many subjects have zero observations. However, many applications have much larger observations, so that the overall mean is not near zero. Thus, the VRU traffic fatal accidents in Rome were not statistically analyzed.

CONCLUSIONS

The use of spatial analysis (KDE method in a GIS environment) presented satisfactory results to identify and delimit zones with high concentrations of VRU traffic accidents. Using the black spots zones, it was possible to classify subzones with different degrees of density (very high/high/medium/low density).

When analyzing the obtained results from São Paulo, the areas with the highest VRU traffic accident rates corresponded with the areas with the highest concentrations of TGH, including hospitals, universities and colleges and retail shopping centers. Figure 9 shows that the area around the hospitals presents the highest risk of VRU traffic accidents. Similarly, Figure 10 illustrates that "very high-density", "high-density" and "medium-density" black spot zones were mainly where universities and colleges are concentrated. The same result can be noticed regarding retail shopping centers (Figure 11).

The statistical analysis of São Paulo revealed an association between the dependent variable (number of VRU total/fatal traffic accidents divided by area) and the independent variables (the ratio trips/inhabitants and the presence of TGH). In turn, the regression model demonstrated that the presence of TGH does not decisively influence the occurrence of fatal VRU traffic accidents, as observed for total accidents (sum of non-fatal and fatal accidents). In terms of fatal VRU accidents, the explanatory variable that presented the highest association was the number of generated trips (trips/inhabitants). In both cases (total and fatal accidents), the OLS model is appropriate for describing the dynamics of VRU traffic accidents and the characteristics of the city (area, population, income, generated trips, TGH, etc.).

According to the visual analysis in Rome, it is not possible to affirm that black spot zones are connected with the preferred destinations of the population (TGH). The statistical analysis showed that the variable "demographic density" is sufficient for explaining the dynamics of total VRU road accidents.

Because Rome is one of the most visited cities in the world and has a large number and variety of historical and touristic attractions that are located in its central area, one possible cause of the high VRU traffic accident rates in these zones could be the high concentration of visitors rather than the presence of traditional trip generator hubs. However, no pattern was observed regarding the TGH. Thus, it is not possible to establish a direct correlation between the presence of trip generator hubs and VRU traffic accident rates in Rome.

FINAL CONSIDERATIONS

According to [76,77], motorized transport modes account for most road accidents in many urban areas. Thus, the development of alternative transport methods, such as mass transit, bicycles and walking, has often been recommended. However, policies directed at such implementations have barely taken off in Rome and São Paulo or for the development of proper safety guidelines for VRU.

Road traffic injuries in Rome are the principal cause of death among young people (14–30 years of age). Managing traffic speed remains crucial for creating a safe road system and for achieving the European Union (EU)-imposed target of reducing road fatalities by 50% in 2011–2020 [78,79].

Similarly, traffic accidents are the main cause of injury and fatalities among children (above one year old) and teenagers (up to 19 years old) in São Paulo [80]. According to [81,82], the number of traffic accidents is related to fast and uncontrolled urban growth in São Paulo. The proportion of fatal crashes involving VRU partly results from cultural and sociodemographic factors. Many fatal traffic accidents or accidents with severe injuries are caused by poor traffic safety conditions, which are often accentuated by physical, political, technical and enforcement environments.

Speed enforcement systems should be adopted to reduce urban road accidents, and the geometric characteristics of the road network should be re-evaluated and modified [83]. One method for reaching these goals is to introduce cost-effective practices, such as traffic calming strategies that benefit the mobility of VRU by reducing the speed of motorized traffic, re-designing transport infrastructure and introducing cycling and walking facilities [84].

Traffic calming refers to a combination of physical changes in road design and speed management that is aimed at improving road safety conditions, especially for users of non-motorized modes (VRU) [85]. Traffic calming measures influence road safety by reducing vehicle speed and/or the volumes of traffic on urban road systems, reducing and/or eliminating conflicting movements, improving visibility, reducing exposure and sharpening drivers' alertness [86]. Traffic calming measures include chicanes, central islands, traffic control devices (e.g., variable message signs or speed cameras), surface treatments (e.g., speed humps or transverse rumble strips) and roadside features (e.g., gateways or landscaping).

The high socio-economic costs of traffic accidents clearly indicate the need for governments and policymakers to strengthen traffic accident preventive measures [87]. According to [78,88], one of the most important intervention methods is educational action, which should be directed at the entire population and especially at adolescents and young adults, which are at higher risk for involvement in traffic accidents. The behaviors of users (drivers and passengers of motorized and non-motorized vehicles and pedestrians) can interfere with the number and severity of traffic accidents, particularly speeding, poor driving skills and education, lack of familiarity with non-motorized transportation modes, lacking the use of safety equipment (seat belt, motorcycle helmet, etc.) and alcohol consumption.

ACKNOWLEDGMENTS

The authors would like to thank the Centro di Ricerca per il Trasporto e la Logistica (Research Centre for Transport and Logistics) at the University of Rome, especially Francesco Filippi, for providing the dataset for Rome and the CET-SP (Companhia de Engenharia de Tráfego de São Paulo, the Traffic Engineering Company of São Paulo) for providing the traffic accident data for São Paulo. In addition, the authors thank the CNPq (Conselho Nacional de Desenvolvimento Científico e Tecnológico, the National Council for Scientific and Technological Development) and FUSP (Fundação da Universidade de São Paulo, the Foundation of the University of São Paulo) for granting scholarships to the researchers and the Laboratory of Geoprocessing in the Department of Transportation Engineering at the Polytechnic School of the University of São Paulo for providing the infrastructure used for this study.

AUTHOR CONTRIBUTIONS

All the authors conceived the study and contributed to the manuscript. C.A.S.M., M.A.G., J.A.Q., A.T., and L.P. performed the spatial analysis. F.C.N. and C.A.S.M. provided the statistical analysis. C.A.S.M. wrote the paper. J.A.Q. and L.P. served as supervisors for this work, respectively in São Paulo (Brazil) and Rome (Italy), and revised the manuscript. All authors read and approved the final manuscript

CONFLICTS OF INTEREST

The authors declare no conflict of interest.

REFERENCES

1. Haque, M.M.; Chin, H.C.; Debnath, A.K. Sustainable, safe, smart—Three key elements of Singapore's evolving transport policies. Transp. Policy **2013**, 27, 20–31.
2. Machado, C.A.S.; Giannotti, M.A.; Shinohara, E.J.; Nishisaki, H.; Quintanilha, J.A. Characterization of the sites of traffic accidents involving vulnerable road users (VRU) in São Paulo city. In Proceedings of the

Transport Research Arena (TRA 2014)—5th Conference Transport Solutions: From Research to Deployment-Innovative Mobility, Mobilise Innovation, Paris, France, 14–17 April 2014; p. 10.

3. Santos, G.; Behrendt, H.; Maconi, L.; Shirvani, T.; Teytelboym, A. Part I: Externalities and economic policies in road transport. Res. Transp. Econ. **2010**, 28, 2–45.

4. Small, K.; Verhoef, E.T. The Economics of Urban Transportation; Routledge: London, UK, 2007; p. 276.

5. Suzuki, H.; Cervero, R.; Iuchi, K. Transforming Cities with Transit—Transit and Land-Use Integration for Sustainable Urban Development; Urban Development Series; The World Bank: Washington, DC, USA, 2013; p. 205.

6. Klassen, J.; El-Basyouny, K.; Islam, M.T. Analyzing the severity of bicycle-motor vehicle collision using spatial mixed logit models: A city of Edmonton case study. Saf. Sci. **2014**, 62, 295–304.

7. Nabors, D.; Goughnour, E.; Thomas, L.; Desantis, W.; Sawyer, M. Bicycle road safety audit guidelines and prompt lists. Federal Highway Administration–FHWA, Final Report; FHWA: Washington, DC, USA, 2012; p. 88.

8. Otte, D.; Jänsch, M.; Haasper, C. Injury protection and accident causation parameters for vulnerable road users based on German In-Depth Accident Study GIDAS. Accid. Anal. Prev.**2012**, 44, 149–153.

9. Schepers, P.; Hagenzieker, M.; Methorst, R.; van Wee, B.; Wegman, F. A conceptual framework for road safety and mobility applied to cycling safety. Accid. Anal. Prev. **2014**, 62, 331–340.

10. Jiang, B.; Okabe, A. Different ways of thinking about street networks and spatial analysis. Geogr. Anal. **2014**, 46, 341–344.

11. Theofilatos, A.; Yannis, G. A review of the effect of traffic and weather characteristics on road safety. Accid. Anal. Prev. **2014**, 72, 244–256.

12. Çela, L.; Shiode, S.; Lipovac, K. Integrating GIS and spatial analytical techniques in an analysis of road traffic accidents in Serbia. Int. J. Traffic Transp. Eng. **2013**, 3, 1–15.

13. Bíl, M.; Andrášik, R.; Janoška, Z. Identification of hazardous road locations of traffic accidents by means of kernel density estimation and cluster significance evaluation. Accid. Anal. Prev.**2013**, 55, 265–273.

14. Wang, C.; Quddus, M.A.; Ison, S.G. The effect of traffic and road characteristics on road safety: A review and future research direction. Saf. Sci. **2013**, 57, 264–275.

15. World Health Organization (WHO). World Report on Road Traffic Injury Prevention; Peden, M., Scurfield, R., Sleet, D., Mohan, D., Hyder, A.A., Jarawan, E., Mathers, C., Eds.; WHO: Geneva, Switzerland, 2004; p. 217.

16. Moeinaddini, M.; Asadi-Shekari, Z.; Shah, M.Z. The relationship between urban street networks and the number of transport fatalities at the city level. Saf. Sci. **2014**, 62, 114–120.

17. World Health Organization (WHO). Global Status Report on Road Safety 2013—Supporting a Decade of Action; WHO: Geneva, Switzerland, 2013; p. 318.

18. Ministry of Healthy. 2013. Available online: http://www2.datasus.gov.br/DATASUS/index.php?area=0205&VObj =http://tabnet.datasus.gov.br/cgi/deftohtm.exe?sim/cnv/ext10 (accessed on 6 May 2013).

19. Koh, P.P.; Wong, Y.D. Comparing pedestrians' needs and behaviours in different land use environments. J. Transp. Geogr. **2013**, 26, 43–50.

20. Flahaut, B.; Mouchart, M.; San Martin, E.; Thomas, I. The local spatial autocorrelation and the kernel method for identifying black zones—A comparative approach. Accid. Anal. Prev. **2003**, 35, 991–1004.

21. Abdel-Aty, M.; Chundi, S.S.; Lee, C. Geo-spatial and log-linear analysis of pedestrian and bicyclist crashes involving school-aged children. J. Saf. Res. **2007**, 38, 571–579.

22. Blazquez, C.A.; Celis, M.S. A spatial and temporal analysis of child pedestrian crashes in Santiago, Chile. Accid. Anal. Prev. **2013**, 50, 304–311.

23. Wang, Y.; Kockelman, K.M. A Poisson-lognormal conditional-autoregressive model for multivariate spatial analysis of pedestrian crash counts across neighborhoods. Accid. Anal. Prev.**2013**, 60, 71–84.

24. Xie, Z.; Yan, J. Detecting traffic accident clusters with network kernel density estimation and local spatial statistics: An integrated approach. J. Transp. Geogr. **2013**, 31, 64–71.

25. Amin, S.R.; Zareie, A.; Amador-Jiménez, L.E. Climate change modeling and the weather-related road accidents in Canada. Transp. Res. Part D: Transp. Environ. **2014**, 32, 171–183.

26. Andrade, L.; Vissoci, J.R.N.; Rodrigues, C.G.; Finato, K.; Carvalho, E.; Pietrobon, R.; Souza, E.M.; Nihei, O.K.; Lynch, C.; Carvalho, M.D.B. Brazilian road traffic fatalities: A spatial and environmental analysis. PLoS ONE **2014**, 9, 1–10.

27. Soltani, A.; Askari, S. Analysis of intra-urban traffic accidents using spatiotemporal visualization techniques. Transp. Telecommun. J. **2014**, 15, 227–232.

28. Vandenbulcke, G.; Thomas, I.; Panis, L.I. Predicting cycling accident risk in Brussels: A spatial case-control approach. Accid. Anal. Prev. **2014**, 62, 341–357.

29. Yu, H.; Liu, P.; Chen, J.; Wang, H. Comparative analysis of the spatial analysis methods for hotspot identification. Accid. Anal. Prev. **2014**, 66, 80–88.

30. Habibovic, A.; Davidsson, J. Causation mechanisms in car-to-vulnerable road user crashes: Implications for active safety systems. Accid. Anal. Prev. **2012**, 49, 493–500.

31. Habibovic, A.; Davidsson, J. Requirements of a system to reduce car-to-vulnerable road user crashes in urban intersections. Accid. Anal. Prev. **2011**, 43, 1570–1580.

32. Anderson, T.K. Kernel density estimation and K-means clustering to profile road accident hotspots. Accid. Anal. Prev. **2009**, 41, 359–364.

33. Ahmed, M.M.; Abdel-Aty, M. Evaluation and spatial analysis of automated red-light running enforcement cameras. Transp. Res. Part C: Emerg. Technol. **2015**, 50, 130–140.

34. Durduran, S.S. A decision making system to automatic recognize of traffic accidents on the basis of a GIS platform. Expert Syst. Appl. **2010**, 37, 7729–7736.

35. Erdogan, S.; Yilmaz, I.; Baybura, T.; Gullu, M. Geographical information systems aided traffic accident analysis system case study: City of Afyonkarahisar. Accid. Anal. Prev. **2008**, 40, 174–181.

36. Kingham, S.; Sabel, C.E.; Bartie, P. The impact of the "school run" on road traffic accidents: A spatio-temporal analysis. J. Transp. Geogr. **2011**, 19, 705–711.

37. Steenberghen, T.; Aerts, K.; Thomas, I. Spatial clustering of events on a network. J. Transp. Geogr. **2010**, 18, 411–418.

38. Xiao, J.; Liu, Y. Traffic incident detection using multiple-kernel support vector machine. J. Transp. Res. Board **2012**, 2324, 44–52.

39. Xie, Z.; Yan, J. Kernel density estimation of traffic accidents in a network space. Comput. Environ. Urban Syst. **2008**, 32, 396–406.

40. Yalcin, G.; Duzgun, H.S. Spatial analysis of two-wheeled vehicles traffic crashes: Osmaniye in Turkey. KSCE J. Civ. Eng. 2015.

41. Brunsdon, C. Estimating probability surfaces for geographical point data: An adaptive kernel algorithm. Comput. Geosci. **1995**, 21, 877–894.

42. Xie, X.; Wu, J. Some improvement on convergence rates of kernel density estimator. Appl. Math.**2014**, 5, 1684–1696.

43. Silverman, B.W. Density Estimation for Statistics and Data Analysis—Monographs on Statistics and Applied Probability, 1st ed.; Chapman and Hall: London, UK, 1986; p. 200.

44. Wand, M.P.; Jones, M.C. Kernel Smoothing—Monographs on Statistics and Applied Probability, 1st ed.; Chapman and Hall: London, UK, 1995; p. 212.

45. Steenberghen, T.; Dufays, T.; Thomas, I.; Flahaut, B. Intra-urban location and clustering of road accidents using GIS: A Belgian example. Int. J. Geogr. Int. Sci. **2004**, 18, 169–181.

46. Ahmed, A.; Khan, B.A.; Khurshid, M.B.; Khan, M.B.; Waheed, A. Estimating national road crash fatalities using aggregate data. Int. J. Inj. Control Saf. Promot. 2015.

47. Bjørnskau, T.; Nævestad, T.-O.; Akhtar, J. Traffic safety among motorcyclists in Norway: A study of subgroups and risk factors. Accid. Anal. Prev. **2012**, 49, 50–57.

48. Chiou, Y.-C.; Fu, C. Modeling crash frequency and severity using multinomial-generalized Poisson model with error components. Accid. Anal. Prev. **2013**, 50, 73–82.

49. Kaplan, S.; Prato, C.G. Risk factors associated with bus accident severity in the United States: A generalized ordered logit model. J. Saf. Res. **2012**, 43, 171–180.

50. Nazif-Munoz, J.I.; Quesnel-Vallée, A.; van den Berg, A. Did Chile's traffic law reform push police enforcement?—Understanding Chile's traffic fatalities and injuries reduction. Inj. Prev.2014.

51. Nóbrega, L.M.; Cavalcante, G.M.S.; Lima, M.M.S.M.; Madruga, R.C.R.; Jorge, M.L.R.; d'avila, S. Prevalence of facial trauma and associated factors in victims of road traffic accidents. Am. J. Emerg. Med. **2014**, 32, 1382–1386.

52. Philip, P.; Chaufton, C.; Orriols, L.; Lagarde, E.; Amoros, E.; Laumon, B.; Akerstedt, T.; Taillard, J.; Sagaspe, P. Complaints of poor sleep and risk of traffic accidents: A population-based case-control study. PLoS ONE **2014**, 9, e114102.

53. Pirdavani, A.; Bellemans, T.; Brijs, T.; Kochan, B.; Wets, G. Assessing the road safety impacts of a teleworking policy by means of geographically weighted regression method. J. Transp. Geogr. **2014**, 39, 96–110.

54. Pirdavani, A.; Bellemans, T.; Brijs, T.; Wets, G. Application of geographically weighted regression technique in spatial analysis of fatal and injury crashes. J. Transp. Eng. 2014.

55. Rangel, T.; Vassallo, J.M.; Arenas, B. Effectiveness of safety-based incentives in public private partnerships: Evidence from the case of Spain. Transp. Res. Part A: Policy Pract. **2012**, 46, 1166–1176.

56. Yao, S.; Loo, B.P.Y.; Lam, W.W.Y. Measures of activity-based pedestrian exposure to the risk of vehicle-pedestrian collisions: Space-time path vs. potential path tree methods. Accid. Anal. Prev.**2015**, 75, 320–332.

57. Agbelie, B.R.D.K. An empirical analysis of three econometric frameworks for evaluating economic impacts of transportation infrastructure expenditures across countries. Transp. Policy2014, 35, 304–310.

58. Qian, X.; Ukkusuri, S.V. Spatial variation of the urban taxi ridership using GPS data. Appl. Geogr. **2015**, 59, 31–42.

59. Wang, Y.; Potoglou, D.; Orford, S.; Gong, Y. Bus stop, property price and land value tax: A multilevel hedonic analysis with quantile calibration. Land Use Policy **2015**, 42, 381–391.

60. Chakrabortya, A.; Mishra, S. Land use and transit ridership connections: Implications for state-level planning agencies. Land Use Policy **2013**, 30, 458–469.

61. Lambert, D.M.; Brown, J.P.; Florax, R.J.G.M. A two-step estimator for a spatial lag model of counts: Theory, small sample performance and an application. Reg. Sci. Urban Econ. **2010**, 40, 241–252.

62. Frondoni, R.; Mollo, B.; Capotorti, G. A landscape analysis of land cover change in the Municipality of Rome (Italy): Spatio-temporal characteristics and ecological implications of land cover transitions from 1954 to 2001. Landsc. Urban Plan. **2011**, 100, 117–128.

63. Zuur, A.F.; Ieno, E.N.; Smith, G.M. Analysing Ecological Data, 26th ed.; Springer: New York, NY, USA, 2007; p. 672.

64. Zhang, L.; Gove, J.H.; Heath, L.S. Spatial residual analysis of six modeling techniques. Ecol. Model. **2005**, 186, 154–177.

65. Razali, N.M.; Wah, Y.B. Power comparisons of Shapiro-Wilk, Kolmogorov-Smirnov, Lilliefors and Anderson-Darling tests. J. Stat. Model. Anal. **2011**, 2, 21–33.

66. Lilliefors, H.W. On the Kolmogorov-Smirnov test for normality with mean and variance unknown.J. Am. Stat. Assoc. **1967**, 62, 399–402.

67. Duncan, D.T.; Kawachi, I.; Kum, S.; Aldstadt, J.; Piras, G.; Matthews, S.A.; Arbia, G.; Castro, M.C.; White, K.; Williams, D.R. A spatially explicit approach to the study of socio-demographic inequality in the spatial distribution of trees across boston neighborhoods. Spat. Demogr. **2014**, 2, 1–29.

68. Boots, B.; Tiefelsdorf, M. Global and local spatial autocorrelation in bounded regular tessellations. J. Geogr. Syst. **2000**, 2, 319–348.

69. Lauridsen, J.; Kosfeld, R. A test strategy for spurious spatial regression, spatial nonstationarity, and spatial cointegration. Pap. Reg. Sci. **2006**, 85, 363–377.

70. Anselin, L. Lagrange multiplier test diagnostics for spatial dependence and spatial heterogeneity.Geogr. Anal. **1988**, 20, 1–17.

71. Akaike, H. Information theory and an extension of the maximum likelihood principle. In Proceedings of the Second International Symposium on Information Theory; Petrov, B., Csáki, F., Eds.; Akadémiai Kiadó: Budapest, Hungary, 1973; pp. 267–281.

72. Dirick, L.; Claeskens, G.; Baesens, B. An Akaike information criterion for multiple event mixture cure models. Eur. J. Oper. Res. **2015**, 241, 449–457.

73. Symonds, M.R.E.; Moussalli, A. A brief guide to model selection, multimodel inference and model averaging in behavioural ecology using Akaike's information criterion. Behav. Ecol. Sociobiol. **2011**, 65, 13–21.

74. Cheung, Y.B. Zero-inflated models for regression analysis of count data: A study of growth and development. Stat. Med. **2002**, 21, 1461–1469.

75. Min, Y.; Agresti, A. Random effect models for repeated measures of zero-inflated count data.Stat. Model. **2005**, 5, 1–19.

76. European Commission. Towards Low Carbon Transport in Europe-Communicating Transporet Research an Innovation; European Union, Directorate General for Mobility and Transport: Brussels, Belgium, 2012; p. 24.

77. Passafaro, P.; Rimano, A.; Piccini, M.P.; Metastasio, R.; Gambardella, V.; Gullace, G.; Lettieri, C. The bicycle and the city: Desires and emotions versus attitudes, habits and norms. J. Environ. Psychol. **2014**, 38, 76–83.

78. Camilloni, L.; Farchi, S.; Chini, F.; Rossi, P.G.; Borgia, P.; Guasticchi, G. How socioeconomic status influences road traffic injuries and home injuries in Rome. Int. J. Inj. Control Saf. Promot.**2013**, 20, 134–143.

79. European Transport Safety Council (ETSC). A Challenging Start towards the E.U. 2020 Road Safety Target; 6th Road Safety PIN Report. ETSC: Brussels, Belgium, 2012; p. 96.

80. Gorios, C.; Souza, R.M.; Gerolla, V.; Maso, B.; Rodrigues, C.L.; Armond, J.E. Acidentes de transporte de crianças e adolescentes em serviço de emergência de hospital de ensino, Zona Sul da cidade de São Paulo. Rev. Bras. Ortop. **2014**, 49, 391–395.

81. Vasconcellos, E.A. Reassessing traffic accidents in developing countries. Transp. Policy **1995**, 2, 263–269.

82. Vasconcellos, E.A. Urban development and traffic accidents in Brazil. Accid. Anal. Prev. **1999**,31, 319–328.

83. Bassani, M.; Dalmazzo, D.; Marinelli, G.; Cirillo, C. The effects of road geometrics and traffic regulations on driver-preferred speeds in northern Italy. An exploratory analysis. Transp. Res. Part F: Traffic Psychol. Behav. **2014**, 25, 10–26.

84. Wegman, F.; Zhang, F.; Dijkstra, A. How to make more cycling good for road safety? Accid. Anal. Prev. **2012**, 44, 19–29.

85. Ghafghazi, G.; Hatzopoulou, M. Simulating the environmental effects of isolated and area-wide traffic calming schemes using traffic simulation and microscopic emission modeling.Transportation **2014**, 41, 633–649.

86. Ariën, C.; Brijs, K.; Brijs, T.; Ceulemans, W.; Vanroelen, G.; Jongen, E.M.M.; Daniels, S.; Wets, G. Does the effect of traffic calming measures endure over time?—A simulator study on the influence of gates. Transp. Res. Part F: Traffic Psychol. Behav. **2014**, 22, 63–75.

87. Alemany, R.; Ayuso, M.; Guillén, M. Impact of road traffic injuries on disability rates and long-term care costs in Spain. Accid. Anal. Prev. **2013**, 60, 95–102.

88. Salvarani, C.P.; Colli, B.O.; Carlotti Júnior, C.G. Impact of a program for the prevention of traffic accidents in a southern Brazilian city: A model for implementation in a developing country.Surg. Neurol. **2009**, 72, 6–14.

CITATION

Cláudia A. Soares Machado, Mariana Abrantes Giannotti , Francisco Chiaravalloti Neto, Antonino Tripodi, Luca Persia and José Alberto Quintanilha, Characterization of Black Spot Zones for Vulnerable Road Users in São Paulo (Brazil) and Rome (Italy), doi:10.3390/ijgi4020858.

CHAPTER 10

Comprehensive Assessment on Sustainable Development of Highway Transportation Capacity Based on Entropy Weight and TOPSIS

Yancang Li [†], Lei Zhao [†] and Juanjuan Suo [*]

College of Civil Engineering, Hebei University of Engineering, Handan 056038, China;
E-Mails: liyancang@hebeu.edu.cn (Y.L.); leizhao@hebeu.edu.cn (L.Z.)

ABSTRACT

With the rapid development of the national economy of China, an increasing need for transportation facilities is becoming a serious challenge that the existing traffic system has to meet. Thus, the highway transportation capacity development level assessment has important significance in theory and in practice. In order to overcome the current defects of stronger subjectivity and experience in common assessment methods, the entropy weight and the TOPSIS method were introduced and employed to the comprehensive assessment of highway transportation capacity development. Shannon information entropy was applied to determine the weight value of each index in the comprehensive assessment model. After determining the index weight, the result of comprehensive assessment was obtained through the TOPSIS method. Finally, the effectiveness and feasibility of the proposed method were shown by application in practice.

INTRODUCTION

With the rapid development of the national economy of China, an increasing need for transportation facilities is becoming a serious challenge that the existing traffic system has to meet. Thus, improving traffic capacity has become a vital way of solving traffic problems. Highway transportation plays an important role in today's economic and social life [1,2,3,4]. It greatly improves and enriches transportation capacity of passengers and cargo due to its strong capacity, high speed and flexible transportation mode [5]. Highway transportation development has a huge impact on economical and social development and becomes a basic condition of various kinds of industry development. Transportation occupies an extremely important position in economic and social development. It should be coordinated with the development of economic and social systems to achieve huge economic and social benefits [6,7,8,9,10,11,12]. At the same time, highway development is also needed intensively when the economy and society develop rapidly in China. A rapid, convenient, comfortable and sustainable transportation network needs to be established urgently alongside economic and social development. The development level shows a developmental status at a time and it reflects the development scale and degree of social economic phenomenon in each period. It is actually one of the specific values in the time series. Through the assessment of the development level of each period, we can know the sustainable development status in all periods and the trend of the development in future, which relates to the sustainability. However, highway transportation is still at a developmental stage in China and a complete and mature transportation system is not established. This will directly restrict the stable development of the Chinese economy and society. So, how best to plan and construct a highway transportation network and realize sustainable development of highway traffic becomes an urgent problem to solve. The comprehensive assessment of highway transportation development becomes an important basis, which evaluates the developmental the pros and cons of an existing highway transportation system. It is playing a vital role in highway transportation development [13]. Currently, many methods and models are applied in the comprehensive assessment of highway transportation development. However, most methods rely on stronger subjectivity and experiences in determining weight value. The objectivity and accuracy of weight value was unavoidably influenced. Thus, the comprehensive assessment results from different researchers are always full of uncertainty. Therefore, we have to find an effective comprehensive method for the assessment of highway transportation development, which has the characteristics of objectivity, simple operation and little interference.

Here, the entropy and the TOPSIS method were employed to the comprehensive assessment of highway transportation development. Shannon information entropy, which is an objective and applicable method for the determination of weight value, was introduced into the comprehensive assessment. It can calculate the weight value of each index more effectively in the comprehensive assessment of highway transportation development. In the application of the Shannon information entropy method, the greater entropy weight indicates a greater variation extent of the relevant index, enabling much more information and having a greater effect. So, the weight value of the corresponding index should also be bigger. In contrast, for the smaller entropy weight, which has little effect, its weight value should be smaller. One of the important contributions of our work is the combined use of entropy weight and TOPSIS, because using only one of these two techniques without the other might lead to unacceptable results.

The structure of this paper is as follows: we first briefly introduce the importance of highway transportation development assessment and clearly point out the advantages of the method employed. Then, the comprehensive assessment model of highway transportation development based on entropy and the TOPSIS method was introduced. In the following part, we focus on the application of the model in practice. Finally, through the application in highway transportation development assessment, the effectiveness and feasibility of the method are shown.

ENTROPY WEIGHT MODEL

Entropy used to be a thermodynamic concept: it was introduced into information theory in 1948 by C. E. Shannon who put forward the concept of information entropy to measure the level of system chaos or disorder [14,15,16,17,18,19]. Shannon information entropy, which is an objective and applicable method for the determination of weight value, was introduced into the comprehensive assessment. In the application of the

Shannon information entropy method, the greater entropy weight indicates greater variation extent of relevant index, which enables much more information and has a greater effect. So, the weight value of the corresponding index should also be bigger. In contrast, for the smaller entropy weight, which has little effect, its weight value should be smaller [17,18].

Data Standardization

Suppose the research plan is x_{ij} ($i = 1,2,\cdots n; j = 1,2,\cdots m$). It denotes that there are i schemes and j indexes in the research plan. Based on appraisal target characteristics, the indexes are divided into the benefit type, the cost type and stationary indexes. The benefit type indexes are the indexes for which bigger values are better. The cost type indexes refer to the indexes for which smaller values are better. The stationary type indexes are the indexes whose values are constant.

For the benefit type indexes, the standardization is as follows:

$$y_{ij} = \frac{x_{ij} - \min x_{ij}}{\max x_{ij} - \min x_{ij}} \tag{1}$$

For the cost type indexes, the standardization is:

$$y_{ij} = \frac{\max x_{ij} - x_{ij}}{\max x_{ij} - \min x_{ij}} \tag{2}$$

where, $\max x_{ij}$ and $\min x_{ij}$ are the maximum and minimum value in index j respectively.

For the stationary factors, the standardization is:

$$y_{ij} = 1 - \frac{x_{ij} - x_{ij}^*}{\max\left|x_{ij} - x_{ij}^*\right|} \tag{3}$$

In the formula, x_{ij}^* is the best stable value in index j.

After the standardized processing, the standard matrix $y = (y_{ij})_{n \times m}$ can be obtained.

Determination of Entropy Weight

Suppose the weight vector of targets is:

$$w = (w_1, w_2, \cdots, w_m), \ 0 \le w_j \le 1 \tag{4}$$

In order to determine weight values effectively, Shannon information entropy is introduced:

$$H = -\sum_{J=1}^{M} w_j \ln w_j \tag{5}$$

Through the formula of information entropy, it is possible to demonstrate that it is a double plan question. It should be transferred to a single objective mathematics model to make it easily calculable. Then, the established single objective mathematics model is as follows:

$$\min u \sum_{i=1}^{n} w_j (1 - y_{ij}) + (1 - u) \sum_{j=1}^{m} w_j \ln w_j \tag{6}$$

where, u is the equilibrium coefficient between two goals, and $0 < u < 1$.

Then the Lagrangian function was employed to solve this model. The Lagrangian function based on (5) is as follows:

$$L(w, \lambda) = u \sum_{i=1}^{n} \sum_{j=1}^{m} w_j (1 - y_{ij}) + (1 - u) \sum_{j=1}^{m} w_j \ln w_j - \lambda (\sum_{j=1}^{m} w_j - 1) \tag{7}$$

After solving (7) based on necessary conditions of existing extreme values, the weight model can be obtained:

$$w_j = \frac{\exp\{-[1 + u \sum_{i=1}^{n} (1 - y_{ij})/(1 - u)]\}}{\sum_{j=1}^{m} \exp\{-[1 + u \sum_{i=1}^{n} (1 - y_{ij})/(1 - u)]\}} \tag{8}$$

TOPSIS MODEL

The TOPSIS method (Technique for Order Preference by Similarity to an Ideal Solution) was first proposed by C.L. Hwang and K. Yoon in 1981. It is an ordering method based on the degree of closeness between limited appraisal objects and idealization objects, and it is a kind of relative advantages and disadvantages evaluation in the existing appraisal objects. It is also called the "advantages and disadvantages distance method". It is an effective method in multi-objective decision analysis [20,21,22,23,24,25]. Its basic principle is to order by calculating distance between appraisal objects and the optimal and worst solutions. If the appraisal objects are close to the optimal solutions, it keeps them away from the worst solutions, which is the best situation.

The ideal point and anti-ideal point are two basic concepts in the TOPSIS method. The ideal point is supposed as the optimal solution, all attribute values of which achieve the best value in all schemes. The anti-ideal point is supposed as the worst solution, all attribute values of which achieve the worst value [24,25].

Supposeis d_i^* the distance between the scheme point and the ideal point, and d_i^- is the distance between the scheme point and anti-ideal point. According to the formula of Hamming distance, d_i^* and d_i^- can be solved as follows:

$$d_i^+ = \sum_{j=1}^{m} w_j (y_{ij} - y_{+j})^{\frac{1}{2}} = \sum_{j=1}^{m} w_j (1 - y_{ij}) \tag{9}$$

$$d_i^- = \sum_{j=1}^{m} w_j (y_{ij} - y_{-j})^{\frac{1}{2}} = \sum_{j=1}^{m} w_j y_{ij} \tag{10}$$

where, y_{+j} is maximum value of index j among n assessment schemes and

$$y_{+j} = \max(y_{1j}, y_{2j}, \cdots, y_{nj}) \tag{11}$$

In contrast, y_{-y} is the minimal value of index j among n assessment schemes.

$$y_{-j} = \min(y_{1j}, y_{2j}, \cdots, y_{nj}) \tag{12}$$

In order to compare the distance between the scheme point and double base points, we can define:

$$T_i = d_i^+ / d_i^-$$
(13)

where, T_i is the relative quality level of the scheme i.

APPLICATION IN HIGHWAY TRANSPORTATION DEVELOPMENT ASSESSMENT

The comprehensive assessment of highway transportation development becomes an important basis, which evaluates the pros and cons of existing highway transportation systems. It plays a vital role in highway transportation development. In order to show the possibility and effectiveness of the proposed method, the highway transportation development of Luoding City in recent years was viewed as an example. Then the comprehensive assessment model was employed in highway transportation development assessment [26]. Here, highway total mileage (X_1), the proportion of city township highway in total mileage (X_2), highway freight turnover (X_3), highway passenger turnover (X_4), motor vehicle population (X_5), highway traffic total investment (X_6), annual average quality rate of highway (X_7), annual daily traffic of national and provincial highway (X_8), and the rate of cement highway to administrative villages (X_9) were selected as the comprehensive assessment indexes according to the specific conditions of Luoding city traffic development. They are all related to sustainability of highway transportation development. The comprehensive assessment indexes of highway transportation development of Luoding City from 2006 to 2010 were shown in Table 1.

Table 1. Comprehensive assessment indexes of highway transportation development of Luoding City.

Assessment index	Years				
	2006	2007	2008	2009	2010
X_1 (km)	1353.3	1357.4	1360.2	1371.5	1854.12
X_2 (%)	83.2	83.2	83.3	83.4	87.7
X_3 (10k ton-km)	36,401	36,488	36,023	37,877	39,226
X_4 (10k passenger-km)	120,895	121,462	114,297	116,415	121,180
X_5 (veh)	5432	6218	6330	6502	9341
X_6 (ten thousand Yuan)	104,710	110,252	103,205	127,854	134,373
X_7 (%)	82.1	90.9	91.5	91.5	92.7
X_8 (veh per day)	39,248	29,855	29,465	31,056	33,891
X_9 (%)	95.8	100	100	100	100

veh: vehicles

The policy-making matrix is:

$$x_{5\times9} = \begin{cases} 1,353.3 & 83.2 & 36,401 & 120,895 & 5,432 & 104,710 & 82.1 & 39,248 & 95.8 \\ 1,357.4 & 83.2 & 36,488 & 121,462 & 6,218 & 110,252 & 90.9 & 29,855 & 100 \\ 1,360.2 & 83.3 & 36,023 & 114,297 & 6,330 & 103,205 & 91.5 & 29,465 & 100 \\ 1,371.5 & 83.4 & 37,877 & 116,415 & 6,520 & 127,854 & 91.5 & 31,056 & 100 \\ 1,854.12 & 87.7 & 39,226 & 121,180 & 9,341 & 134,373 & 92.7 & 33,891 & 100 \end{cases}$$

Then, the standard matrix $y = (y_{ij})_{5\times9}$ can be obtained based on policy-making matrix:

$$y_{5\times9} = \begin{cases} 0 & 0 & 0.1180 & 0.9209 & 0 & 0.0483 & 0 & 1 & 0 \\ 0.0082 & 0 & 0.1452 & 1 & 0.2011 & 0.2261 & 0.8302 & 0.0399 & 1 \\ 0.0134 & 0.0222 & 0 & 0 & 0.2297 & 0 & 0.8868 & 0 & 1 \\ 0.0363 & 0.0444 & 0.5788 & 0.2956 & 0.2783 & 0.7908 & 0.8868 & 0.1626 & 1 \\ 1 & 1 & 1 & 0.9606 & 1 & 1 & 1 & 0.4524 & 1 \end{cases}$$

Here, suppose $u = 0.5$, according to the formula (8), the weight vector of m targets can be obtained. The weight value of each index is shown in Table 2.

Table 2. Weight value of each index.

Assessment index	Weight value
X_1	0.0196
X_2	0.0199
X_3	0.0431
X_4	0.1641
X_5	0.0379
X_6	0.0539
X_7	0.2517
X_8	0.0359
X_9	0.3739

Then, based on the TOPSIS model, the distances between scheme point and double base points can be determined. Relative quality level of the scheme i T_i was shown in Table 3.

Table 3. Relative quality level.

T_i	Value
T_1	4.1361
T_2	0.2910
T_3	0.6488
T_4	0.3676
T_5	0.0268

We can obtain the result of comprehensive assessment of highway transportation development of Luoding City according to the table above. The result of comprehensive assessment was shown in Table 4.

Table 4. Result of comprehensive assessment.

Years	Order of development level
2006	5th
2007	2nd
2008	4th
2009	3rd
2010	1st

From Table 4, we can know the highway transportation development conditions of Luoding City in recent years. From 2006 to 2008, the highway transportation development level gradually increased. However, the highway transportation development was lower than the previous year. It simply reflected the actual development situation and conformed the development environment at that time. Since 2009, the highway transportation development level has gradually increased. Moreover, the development level is relatively high. Here, the results of highway transportation development assessment we obtained were in accordance with the results of [21]. The effectiveness and feasibility of the proposed method were shown.

CONCLUSIONS

The comprehensive assessment of highway transportation development level has important significance in theory and practice. It is also an important basic for realizing the status of highway transportation capacity development. Here, entropy and the TOPSIS method were employed to evaluate objectively and effectively the status of highway transportation capacity development. From the analysis of the assessment indexes, we know that the highway passenger turnover, annual average quality rate of highway, and the rate of cement highway to administrative villages are the three primary factors with higher weight coefficient. Moreover, more attention should be paid to them in future highway traffic development. From the final comprehensive assessment result obtained, we can also assess the highway traffic development status and trend of Luoding city in recent years. The results show that highway traffic of Luoding city has an improved development status. Moreover, with the rapid development of the economy and the increasing investment in highway construction, we can anticipate further positive development in the future. This assessment method can control and guide the development of the highway transportation capacity further. The comprehensive assessment results are acceptable and they showed the effectiveness and feasibility of the method proposed. The study provides an effective method for the assessment of highway transportation development level and the related fields.

ACKNOWLEDGMENTS

The work was supported by Natural Science Foundation of Hebei Province, China. (No. E2012402030) and the Program of Selection and Cultivating of Disciplinary Talents of Colleges and Universities in Hebei Province (BR2-206).

AUTHOR CONTRIBUTIONS

Yancang Li designed research; Lei zhao performed research and analyzed the data; Juanjuan Suo wrote the paper. All authors read and approved the final manuscript.

CONFLICTS OF INTEREST

The authors declare no conflict of interest.

REFERENCES

1. Amekudzi, A.A.; Khisty, C.J.; Khayesi, M. Using the sustainability footprint model to assess development impacts of transportation systems. *Transport. Res. Pol. Pract.* **2009**, *43*, 339–348.
2. Xiong, Q.; Pu, L.; Yu, M.Y. Analysis of transportation development needs requirement and policy suggestions for road highway transportation in urban agglomeration areas in China. Available online: http://ascelibrary.org/doi/abs/10.1061/9780784413159.179 (accessed on 20 October 2013).
3. Yi, L.X.; Williams, B.M.; Rouphail, N.M.; Chase, R.T. Development of an oversaturated speed-flow model based on the highway capacity Manual. *J. Transport. Res. Board* **2013**, *2395*, 41–48.
4. Li, H.D.; Zhou, Z.Y. Sustainable development of transportation and design of its index system. *Environ. Prot. Transport.* **2003**, *24*, 17–19, (In Chinese).
5. Fricker, J.D.; Whitford, R.K. *Fundamentals of Transportation Engineering*; Prentice Hall: Upper Saddle River, NJ, USA, 2004.
6. Rephann, T.; Isserman, A. New highways as economic development tools: An evaluation using quasi-experimental matching methods. *Reg. Sci. Urban Econ.* **1994**, *24*, 723–751.
7. Rephann, T.J. Highway investment and regional economic development: Decision methods and economical foundations. *Urban Stud.* **1993**, *30*, 437–450.
8. Roh, H.-J.; Datla, S.; Sharma, S. Effect of Snow, Temperature and Their Interaction on Highway Truck Traffic. *J. Transport. Tech.* **2013**, *3*, 24–38.
9. Weisbrod, G. Models to predict the economic development impact of transportation projects: Historical experience and new applications. *Ann. Reg. Sci.* **2008**, *42*, 519–543.
10. Aschauer, D.A. Highway capacity and economic growth. *Econ. Perspect.* **1990**, *14*, 14–24.
11. Shirley, C.; Winston, C. Firm inventory behavior and the returns from highway infrastructure investments. *J. Urban Econ.* **2004**, *55*, 398–415.
12. Berechman, J.; Ozmen, D.; Ozbay, K. Empirical analysis of transportation investment and economic development at state, county and municipality levels. *Transportation* **2006**, *33*, 537–551.
13. Litman, T.; Burwell, D. Issues in sustainable transportation. *Int. J. Global Environ. Issues* **2006**,*6*, 331–347.
14. Liu, P.D.; Han, Z.S. A fuzzy multi-attribute decision-making method under risk with unknown attribute weights. *Technol. Econ. Dev. Econ.* **2011**, *2*, 146–258.

15. Ma, L.H.; Kang, S.Z.; Xuan, Y.Q.; Su, X.L.; Solomatine, D.P. Analysis and simulation of the influencing factors on regional water used based on information entropy. *Water Policy* **2012**, *14*, 1033–1046.

16. Chen, T.Y.; Li, C.H. Determining objective weights with intuitionistic fuzzy entropy measures: A comparative analysis. *Inf. Sci.* **2010**, *180*, 4207–4222.

17. Chen, H.T.; Tsutsumi, M.; Yamasaki, K.; Iwakami, K. An impact analysis of the Taiwan Taoyuan international airport access MRT system considering the interaction between land use and transportation behavior. *J. East. Asia Soc. Transport. Stud.* **2013**, *10*, 315–334.

18. Liu, L.; Zhou, J.Z.; An, X.L.; Zhang, Y.C.; Yang, L. Using fuzzy theory and information entropy for water quality assessment in three Gorges Region, China. *Expert Syst. Appl.* **2010**, *37*, 2517–2521.

19. Calabrese, F.; Diao, M.; di Lorenzo, G.; Ferreira, J.; Ratti, C. Understanding individual mobility patterns from urban sensing data: A mobile phone trace example. *Transport. Res. Emerg. Technol.* **2013**, *26*, 301–313.

20. Li, P.Y.; Wu, J.H.; Qian, H. Groundwater quality assessment based on rough sets attribute seduction and TOPSIS method in a semi-arid area, China. *Environ. Monit. Assess.* **2012**, *184*, 4841–4854.

21. Torlak, G.; Sevkli, M.; Sanal, M.; Zaim, S. Analyzing business competition by using fuzzy TOPSIS method: An example of Turkish domestic airline industry. *Experts Syst. Appl.* **2011**, *38*, 3396–3406.

22. Soltani, A.; Marandi, E.Z.; Ivaki, Y.E. Bus route evaluation using a two-stage hybrid model of Fuzzy AHP and TOPSIS. *J. Transp. Lit.* **2013**, *7*, 34–58.

23. Janic, M.; Reggiani, A. An application of the multiple criteria decision making (MCDM) analysis to the selection of a new hub airport. *Eur. J. Transport Infrastruct. Res.* **2002**, *2*, 113–139.

24. Bu, L.; van Duin, J.H.R.; Wiegmans, B.; Luo, Z.; Yin, C.Z. Selection of city distribution locations in urbanized areas. *Soc. Behav. Sci.* **2012**, *39*, 556–567.

25. Ozceylan, E. A decision support system to compare the transportation modes in logistics. *Int. J. Lean Think.* **2010**, *1*, 58–83.

26. Liu, J.B. *Evaluation Research of Regional Highway Transportation Development Based on Information Entropy*; Press of South China University of Technology: Guangzhou, China, 2011; (In Chinese).

CITATION

Yancang Li , Lei Zhao and Juanjuan Suo, Comprehensive Assessment on Sustainable Development of Highway Transportation Capacity Based on Entropy Weight and TOPSIS, doi:10.3390/su6074685.

CHAPTER 11

Reliable Freestanding Position-Based Routing In Highway Scenarios

Gabriel A. Galaviz-Mosqueda [1], Raúl Aquino-Santos [2], Salvador Villarreal-Reyes [1],, Raúl Rivera-Rodríguez [1], Luis Villaseñor-González [3] and Arthur Edwards [2]*

[1]Electronics and Telecommunications Department, CICESE Research Center, Ensenada-Tijuana Highway, Km. 3918, Playitas, C. P. 22860, Ensenada, Baja California, Mexico;
E-Mails:agalaviz@cicese.edu.mx (G.A.G.-M.); rrivera@cicese.mx (R.R.-R.)
[2]Faculty of Telematics, University of Colima, Av. University 333, C. P. 28040, Colima, Col., Mexico; E-Mails: aquinor@ucol.mx (R.A.-S.); arted@ucol.mx (A.E.)
[3]Plantronics Inc. Mexico, Av. Producción No. 216 Parque Industrial Internacional Tijuana, Mexico; E-Mail: luis.villasenor@plantronics.com

ABSTRACT

Vehicular *Ad Hoc* Networks (VANETs) are considered by car manufacturers and the research community as the enabling technology to radically improve the safety, efficiency and comfort of everyday driving. However, before VANET technology can fulfill all its expected potential, several difficulties must be addressed. One key issue arising when working with VANETs is the complexity of the networking protocols compared to those used by traditional infrastructure networks. Therefore, proper design of the routing strategy becomes a main issue for the effective deployment of VANETs. In this paper, a reliable freestanding position-based routing algorithm (FPBR) for highway scenarios is proposed. For this scenario, several important issues such as the high mobility of vehicles and the propagation conditions may affect the performance of the routing strategy.

These constraints have only been partially addressed in previous proposals. In contrast, the design approach used for developing FPBR considered the constraints imposed by a highway scenario and implements mechanisms to overcome them. FPBR performance is compared to one of the leading protocols for highway scenarios. Performance metrics show that FPBR yields similar results when considering free space propagation conditions, and outperforms the leading protocol when considering a realistic highway path loss model.

INTRODUCTION

In the context of Intelligent Transport Systems (ITS), Vehicular *Ad Hoc* Networks (VANETs) are considered the key technology required to radically improve safety [1,2]. Their importance can also be inferred from the growing interest that VANETs have drawn from entities directly involved in ITS research and development, [3–5]. In fact, according with the National Highway Traffic Safety Administration, VANETs (also called vehicle-to-vehicle communications or V2V) potentially address 79% of all pre-crash scenarios involving unimpaired drivers [1]. Furthermore, it is also expected that a wide range of applications will become available by enabling V2V communication or vehicle-to-infrastructure communications (V2I). Some examples of important applications include safety [6], Internet connectivity [7], entertainment through multiplayer games [8], file sharing support [9], lane change assistance [2] and traffic control [10].

For several years, consortiums and car manufacturers around the World have focused their efforts on enhancing driver safety, comfort and efficiency [11–16]. These efforts have already provided meaningful results, which range from simulation tools [17] to real prototypes of vehicular safety applications [18]. A significant result is a family of international standards that shares a common architecture, networking protocols and air interface definitions for VANET wireless communications. This family of standards is called Communication Access for Land Mobile (CALM) [19], which was adopted by the European community in 2006. Additionally, the Institute of Electrical and Electronic Engineers (IEEE) has significantly contributed to VANET development by providing a series of standards for Dedicated Short-Range Communications (DSRC), such as the IEEE 802.11p standard [20]. The research community has also been actively contributing in different research areas related to the development of VANETs. Examples of active research topics in VANETs include medium access control methods [21], novel applications [8,9], security [22,23] and routing protocols [24–26].

In VANETs the network topology is *ad hoc* in nature. Hence, no infrastructure beyond the network adapter inside the vehicles is necessary. Importantly, the self-dependence of VANETs offers two main advantages: ubiquitous information sharing and low implementation costs. Although there are several communication standards that can be used to establish a radio link between vehicles in a VANET such as mobile WiMAX [27], IEEE 802.11b radios [28–30] and Bluetooth [31–33], currently the IEEE 802.11p standard is emerging as the most prominent option. The IEEE 802.11p standard is based on the physical layer (PHY) of the IEEE 802.11a standard. The main difference is that IEEE 802.11p limits the channel bandwidth to 10 MHz and sets the operating frequency to 5.9 GHz [34]. Importantly, the Medium Access Control (MAC) of IEEE 802.11p is still based on the Carrier Sense Multiple Access with Collision Avoidance (CSMA/CA) method. Despite the advantages offered by CSMA/CA, a significant issue arising when using this MAC mechanism is that it cannot guarantee message dissemination by itself. Another issue with CSMA/CA is that it cannot provide a time frame for packet delivery because of the exponential back off mechanism. Lastly, it is worth mentioning that IEEE 802.11p adapts the enhanced distributed channel access (EDCA) mechanism of the IEEE 802.11e standard (with some modifications to the transmission parameters). Shortly, the EDCA mechanism defines different distributed coordination function (DCF)-CSMA/CA parameters for four different traffic classes: background, best effort, video and voice, [20,35,36].

Currently, the design of routing strategies represents one of the most important research topics for VANETs. In fact, routing is one of the most critical and challenging components that must be solved to enable V2V communications. For example, the routing protocol must deal with the constraints imposed by the high mobility of vehicles. Furthermore, constrains such as mobility patterns, fading wireless channel, density of vehicles and the availability of infrastructure are closely related to the specific deployment scenario, *i.e.*, highway or urban [26,37]. For example, the speeds achievable on highways are significantly higher than in urban scenarios. Thus, a protocol designed to work in urban scenarios may not be able to cope with the higher vehicle speeds present in highway scenarios.

Therefore, the specific application scenario should be considered when designing a routing strategy in order to adequately address the particular constraints imposed by a given scenario. In this context, it is important to note that highways account for a significant amount of the road infrastructure deployed throughout several countries. For example, highways represent about 75% of the total statute miles in the U.S. [1]. Deploying roadside infrastructure to provide coverage for the entire network of roads and highways would take a long time and require a huge investment. Hence, enabling vehicle communications in highways through the use of VANETs is an important scenario that should be further studied, as recently research have clearly pointed out [6,21,38–42].

This paper introduces an efficient multi-hop routing mechanism to enable V2V communications in highway scenarios without employing underlying infrastructure. The routing algorithm was developed considering that the vehicles speed can be as high as 60 m/s under regular traffic conditions. Furthermore, other important constraints pertaining to this application scenario are considered within the routing algorithm design including: dynamic transmission ranges, vehicle acceleration, and high vehicle mobility. Our proposed routing strategy addresses these issues by using a broadcast approach with a light retransmission mechanism in the location service, altogether coupled with a reliable position-based routing algorithm. It is important to note that the routing strategy considers realistic propagation conditions following the results reported in [37]. Our main contributions include:

1. A new location service based on a broadcast approach coupled with a new broadcast suppression technique. The location service addresses the hard conditions imposed by the wireless channel, providing a high success rate for route discoveries.
2. A novel light routing protocol that considers the constraints imposed by a realistic channel propagation model in its design.

The remainder of this paper is organized as follows: Section 2 provides a literature review. Section 3 introduces the Freestanding Position Based Routing (FPBR) protocol proposed in this paper. Section 4 then presents several performance metrics obtained when using FPBR for V2V communications in highway scenarios. The performance evaluation setup is also described in this section. Finally, Section 5 presents the conclusions of this work.

RELATED WORK

The VANETs routing problem was initially addressed by using several well-known routing protocols such as AODV, DSR, and GPSR [43–45], which were originally developed for mobile *ad hoc* networks (MANETs). However, these protocols were not designed to cope with the specific constraints found in VANETs such as dynamic transmission ranges, vehicle acceleration and high vehicle mobility [46]. Consequently, the performance of these protocols for use in VANETs is not as good as that shown for low mobility applications. Subsequent routing protocols, such as those introduced in [47–50], were specifically developed for VANET applications. In [48], the authors propose a junction-based unicast routing algorithm for urban VANETs called GyTAR. This algorithm consists of three main stages: traffic density estimation, intersection selection and data forwarding between intersections. The features of the GyTAR protocol aim to overcome packet losses. Nevertheless, GyTAR assumes the presence of fixed wireless routers at intersections, which may not be easy to deploy in highway scenarios. Another feature of GyTAR is that it assumes that the vehicles' geographic position is provided by an ideal external location service (LS). The drawback with the ideal LS assumption is that the LS effects in the routing metrics are unknown. For the radio channel, GyTAR considers a radio range, based on the two-ray model. Another junction-based routing algorithm for urban scenarios called CMGR was introduced in [51]. In this work, the authors propose an algorithm to discover a route to any gateway attached to a backhaul network. The geographic greedy forwarding algorithm is used by CMGR to relay the packets between intersections. The discovery service of CMGR is similar to route establishment in AODV. Although the broadcast mechanism implemented in CMGR allows fast packet propagation, CMGR does not consider any redundancy strategy. Including a redundancy strategy within the broadcast mechanism is very important because severe wireless channel conditions can cause significant packet losses. The CMGR and GyTAR protocols addressed important challenges commonly found in VANETs. Nevertheless, it can be argued that their use is not the most adequate for highway scenarios as they are infrastructure dependent [41]. Additionally, as shown in [37], the radio channel in highway scenarios does not correspond to the propagation models used in the performance evaluation of these protocols. Another routing protocol developed for urban scenarios named TO-GO is introduced in [50]. TO-GO is a geo-opportunistic routing protocol that requires previous knowledge of the VANET topology in order to select the next forwarder node. For this purpose, each node in TO-GO must construct a 2-hop neighbors table by means of periodic beacon messages. In addition to the next forwarder node, the transmitter node selects a set of backup nodes by means of a complex procedure involving the use of Bloom filters. The chosen forwarder node and the set of backup nodes form a forwarder set that uses a distance-based timer to determine which node will retransmit the packet first. In order to avoid the broadcast storm problem, every node in the forwarder set must be able to hear each other. However, it is important to take into account that the Bloom filter can give false positives in dynamic data sets [52], thus possibly causing unwanted transmissions. Although [50] reports good performance metrics for TO-GO when considering a non-ideal path loss model, the implemented channel was not explicitly developed for V2V scenarios. Finally, it is worth mentioning that TO-GO does not implement a location discovery service. CAR is a routing protocol developed for city and highway scenarios [47]. It includes its own location service, which is based on the Preferred Group Broadcasting technique (PGB [53]). In PGB, every relay node is registered in the broadcast packet if its velocity vector direction is non-parallel to the velocity vector direction of the previous forwarder node. The location service includes a retransmission mechanism, which consists of retransmitting the broadcast packet if the original sender does not hear the transmission of the same packet from a forward neighbor. Although this retransmission mechanism helps overcome the loss of packets, it could also lead to unnecessary retransmissions. Furthermore, route replay is performed by a unicast strategy, which has one main drawback: other nodes may not be warned about the search. This may lead to unnecessary searches, with the consequent waste of resources. The GPSR-L protocol was introduced in [49] with the aim of overcoming some of the problems shown by GPSR when used in VANETs. In GPSR-L, a lifetime threshold for each link is introduced in the neighbors table in order to overcome the lack of a prediction mechanism in GPSR. With the introduction of this threshold, GPSR-L aims to overcome erroneous next-hop selections caused by old information stored in the neighbors table. Nevertheless, for highway scenarios, erroneous next-hop selections caused by the channel can still occur as a deterministic radio channel was assumed in [49]. The

LORA-CBF protocol is a position- based routing algorithm introduced in [54]. This protocol implements a clustering algorithm where only the gateway nodes are allowed to retransmit packets. Nevertheless, since freespace propagation conditions are assumed in LORA-CBF, the clustering algorithm does not guarantee cluster stability in highway scenarios. This may lead to packet losses and/or delays, because it may be difficult to form clusters or an excessive number of clusters may be formed (consisting of a single node) depending on the particular channel conditions at any given moment. Another problem arising with LORA-CBF in highway scenarios is that the most forward within radius neighbor is selected as the next-hop when disseminating data. As it will be explained in the following sections, this assumption may lead to significant packet losses caused by the constraints imposed by the highway V2V channel. In [39], the authors introduced a routing protocol specifically developed for VANETs deployment in highway scenarios. This protocol was named Destination Discovery Oriented Routing (DDOR). Compared to routing protocols previously introduced for VANETs, the DDOR protocol offers advantages such as a built-in location service and a low complexity position prediction algorithm. The inclusion of these mechanisms is important, because they help to overcome erroneous next-hop selections. Additionally, DDOR proposes enabling the IEEE 802.11 RTS/CTS mechanism as a redundancy strategy for its location service and for the data dissemination stages. The LS included in DDOR is an important feature that makes DDOR more suitable for highway scenarios. Nevertheless, one weakness of DDOR is that its redundancy strategy is not controlled by itself. A consequence of this lack of control is that erroneous next-hop selections may lead to a significant overhead increase in the MAC layer. Despite this weakness, it seems that for highway scenarios DDOR overcomes several of the problems shown by the routing protocols previously mentioned. In fact, in [39] several performance metrics were provided for DDOR, showing that this protocol outperforms AODV, DSR and GPSR.

It has been suggested that GPSR-L may offer good performance in highway scenarios compared to CAR, GPSR and GyTAR [26]. However, GPSR-L was not developed with the highway scenario in mind. Furthermore, for this scenario GPSR-L has the drawback that a packet retransmission strategy, a discovery service, and a strategy to handle the acceleration of vehicles were not implemented. The inclusion of these strategies in a routing protocol for highway VANETs is very important, because of the wireless channel constraints imposed by these scenarios may lead to significant packet losses. Furthermore, a self-dependent algorithm is more suitable for this kind of scenarios [26,39]. In contrast, DDOR was explicitly developed for highway scenarios. Therefore, it does include a packet retransmission strategy and a discovery service.

One common characteristic found in AODV, DSR, GPSR, GyTAR, TO-GO, CGMR, LORA-CBF, CAR, GPRS-L and DDOR is that most of them where designed without considering a specific channel model. In fact, most of them have been evaluated by considering an ideal path loss model or channel models that were not specifically developed for V2V communications. Although this assumption is a good starting point for the design and evaluation of a routing strategy, recent works have shown that the path loss model has a significant impact in the routing algorithm performance [55]. Furthermore, the authors in [37] showed that the relative direction of vehicles seriously affects the link path loss in highway scenarios. Therefore, it is really important to consider the path loss model when designing new routing mechanisms for highway scenarios.

Based on the previous discussion, it can be stated that two significant issues that must be addressed when designing new routing protocols for highway VANETs are: the selection of an adequate V2V wireless propagation channel model [37], and the design of a reliable destination discovery service. As explained in Section 3, the FPBR protocol introduced in this paper considers these design constraints.

FREESTANDING POSITION-BASED ROUTING PROTOCOL

As previously explained, the constraints imposed by highway channels must be considered when developing a routing strategy for highway VANETs, this in order to achieve a satisfactory routing performance when deploying VANETs in this kind of scenarios. Therefore, the different modules of the routing protocol (e.g., the beaconing period, the next-hop selection algorithm, the location discovery service, and location prospection) should be designed with the aim to overcome the drawbacks imposed by the highway scenarios. The next subsections introduce the different modules that comprise the Freestanding Position Based Routing (FPBR) protocol proposed in this paper. These modules were specifically developed for highway VANETs. For the development of FPBR it was assumed that the nodes are equipped with digital maps, a global positioning system (GPS), an IEEE 802.11a radio transceiver or an IEEE 802.11p radio transceiver with a single traffic class enabled. It is considered that all the nodes in the VANET use the same radio transceiver (either IEEE 802.11a or IEEE 802.11p). Additionally, a distributed strategy for traffic density estimation like the proposed in [56,57] is also assumed.

Radio Channel

It is well known that in a wireless communication system, the impairments introduced by the channel can adversely affect the transmitted signal. In this context, the path loss and the fading are the most important channel properties to consider when analyzing the performance of a V2V wireless communication system, [37,58,59]. In fact, as the path loss increases the SNR at the receiver decreases, resulting in a degradation of the system performance, [37,58,59]. Most of the research dealing with the design of routing protocols for VANETs assumes that the radio channel follows the behavior of traditional propagation models, e.g., freespace, two rays, Nakagami, *etc.* Nevertheless, as shown in [37], these models do not hold for V2V communications in highway scenarios, because of the antennas height and the vehicle direction have different effects over the path loss [37,55,58,59].

Considering an adequate radio channel model for the development of a routing protocol aimed to V2V applications is very important, as can be inferred from the analysis and results reported in [55]. In fact, the results in [55] show that different radio channel models have different impacts on the routing strategy performance. This is a consequence of the VANETs PHY layer intrinsic features, specifically the low antenna heights and the highly dynamic topology, [55,58–60]. Recently, several channel models for V2V communications have been proposed in the literature, e.g., see [37,61–64] and references within [58,59]. Among these works, [37,62–64] provide path loss models for highway scenarios. On the other hand, [61] provides two different small-scale fading models for two different cases of V2V communications in highways: oncoming vehicles, and vehicles traveling in the same direction at the same speed. The models introduced in [61] have been previously used for the evaluation of the IEEE 802.11p standard, [27]. However, these models are not straightforward applicable to the evaluation and design of routing protocols for VANETs in highway scenarios, since the path loss and the shadowing are not considered. Particularly, the path loss is a critical parameter that should be considered when evaluating VANETs routing protocols, because of the SNR of the received signal at different nodes is dependent on the path loss. Furthermore, [61] does not provide a model for vehicles moving away each other in highways, which is a case occurring in highway VANETs [37].

Among the works that provide path loss characterization for highway scenarios (e.g., [37,62–64]), the model introduced in [37] stands out because it is derived from extensive V2V measurement campaigns. Moreover, this model agrees well with previous findings as those reported in [63]. The path loss model introduced in [37] follows a classic power law and includes a large-scale fading (shadowing) term. Furthermore, the path loss model in [37] allows the characterization of three different V2V communications cases with one single model, specifically: oncoming vehicles, vehicles traveling in convoy, and vehicles travelling away from each other. Therefore, this work considers the highway path loss model introduced in [37] for the development of the routing strategy, since this model was specifically developed for V2V highway scenarios. The path loss model introduced [37] is described byEquation (1):

$$PL(d) = PL_0 + 10nlog_{10}\left(\frac{d}{d_0}\right) + X_\sigma + \zeta PL_c \quad d > d_0$$

$$(1)$$

where d is the propagation ($T_x - R_x$) distance; PL_0 is the path loss at a reference distance d_0; n is the path loss exponent; X_σ is a zero-mean normally distributed random variable with standard deviation σ; PL_c is a correction term that accounts for the offset between forward and reverse path loss. An important characteristic of this path loss model is the introduction of the ζ variable, which is defined according to the relative direction of vehicles. Specifically, ζ is set to: 1 for vehicles travelling in opposite directions and moving away; −1 for vehicles travelling in opposite directions and getting closer; and 0 for vehicles in travelling in convoy. Specific values for every parameter in (1) can be found in [37].

Beacon Rate

The vehicles (nodes) choose the next hop for sending data or control messages based on information previously exchanged with neighbors. This information is exchanged by means of beacon messages. Consequently, there is a tradeoff between the overhead caused by the beacons and the age of the information in the one-hop neighbors table. As different VANET scenarios have different mobility patterns, the strategy to handle this tradeoff depends on the scenario specific characteristics [26]. Thus, setting an adequate beacon rate is an important task, since overly high beacon rates may lead to unnecessary overhead whereas overly low beacon rates may lead to erroneous next hop selections, (e.g., see [65]). Therefore, setting an adequate beacon rate is an important task that depends on the particular routing protocol features. As an example AODV, GyTAR, LORA-CBF and GPSR use a fixed beacon rate of 1 beacon per second, whereas DDOR uses a variable beacon rate ranging from 0.2 to 0.4 beacons per second.

The same set of vehicles is available most of the times in highway scenarios, [26]. Therefore, it may be feasible to use a location prediction (LP) algorithm to achieve a relatively low beacon rate. Consequently, a LP algorithm has been implemented within FPBR in order to predict the gap between the stored and the actual position of neighbor vehicles. This LP algorithm makes possible to achieve beacon rates of 0.4 when considering a free flow vehicle density (this density implies that the vehicles will reach its maximum speed [66,67]), and beacon rates of 0.3 for other vehicle densities.

Next-Hop Selection Algorithm

Location Prediction Algorithm

When a vehicle needs to send or relay a packet, the location prediction algorithm must fill the gap between the last values stored in the one-hop neighbors table and the actual values of each neighbor. Thus, when a next-hop neighbor has to be selected, the LP algorithm is invoked before performing the selection. The LP algorithm implemented in FPBR works as follows:

1. The number of entries in the one-hop neighbors table is updated as follows: if the age of the position information for one particular

neighbor is older than 2 times the beacon rate, then this particular neighbor is deleted from the one-hop neighbors table.

2. After performing step 1, the position of all neighbors is updated in the one-hop neighbors table by means of Equation (2):

$$P_{prospected} = P_{current} + (\hat{v} * dt + \widehat{acc} * dt^2)$$

(2)

where $\hat{v}$ is the last stored velocity vector of the neighbor; $\widehat{acc}$ is the last stored acceleration vector of the neighbor; and dt is the information dwell time in the one-hop neighbors table.

3. The updated one-hop neighbors table is passed to the next-hop selection algorithm.

Selection Algorithm
The FPBR protocol selects the next hop node by first predicting the actual position of each node with the LP algorithm. After prediction, vehicles are grouped in three different sets, based on their relative direction to the sender vehicle: vehicles traveling in opposite directions and moving away belong to group V_s; vehicles traveling in opposite directions and approaching belong to group V_a; and vehicles traveling in the same direction belong to group V_c. The final selection is made considering the most forward within adjusted radio (MFWAR) mechanism introduced in this paper. The adjustment of the radio range coverage is performed considering the radio propagation model introduced in [37] as described byEquation (1).

The Most Forward Within Adjusted Radius (MFWAR) mechanism
Equation (1) shows that it is very important to consider the relative direction of the neighbors when the next forwarder is selected. According to Equation (1), vehicles travelling in opposite directions and approaching have a higher probability of successfully receiving a packet, because of the relative path loss decrease (*i.e.*, ζ is set to -1). When vehicles move away from each other in opposite directions, the probability of dropping a packet becomes higher because of the relative path loss increase (*i.e.*, ζ is set to 1). If the vehicles are traveling in convoy, then there is neither a relative path loss increase nor a relative path loss decrease (*i.e.*, ζ is set to 0). Note that this does not imply that the path loss remains unchanged when the vehicles travel in convoy, since the shadowing effects are still considered

by Equation (1). Furthermore, from Equation (1) it can be inferred that considering the transmitter nominal radio coverage for the selection process may not be the best option, since the shadowing may lead to significant packet losses near the border of the nominal transmission range. Thus, in order to decrease the probability of dropping a packet, the most forward within radius (MFR) technique (see [68] for a description of MFR) was modified to include the use of three dynamic factors named β_a, β_c, and β_s. The dynamic factors are used to scale the nominal radio coverage to a more accurate value for each set of vehicles V_a, V_c, and V_s respectively. We call this new dissemination technique *Most Forward Within Adjusted Radius* (MFWAR).

In order to explain how the dynamic factors are used, assume that node T_1 in Figure 1(a) has nominal radio range R_n, and that β_s and β_c are the scaling factors for the V_s and V_c sets respectively. Additionally the following assumptions are made for this figure:

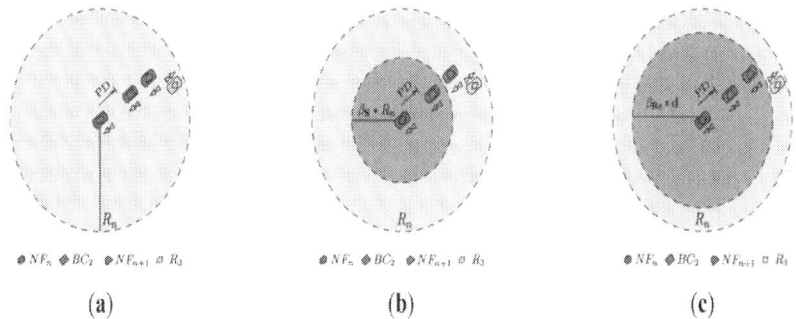

(a) (b) (c)

Figure 1. Use of the dynamic factors β_s. and β_c to adjust the nominal radio range, R_n, for vehicles belonging to the V_s and V_c sets. (a) Nominal radio coverage R_n; (b) Adjusted $\beta_s * R_n$ radio range for vehicles traveling in opposite direction and moving away (V_s set). (c) Adjusted $\beta_c * R_n$ radio range for vehicles traveling in the same direction (V_c set).

- Node T_1 has data to send beyond its nominal radio coverage.
- Node R_3 moves away from T_1 and thus it belongs to the V_s set.
- Nodes R_1 and R_3 travel in the same direction of T_1 and belong to the V_c set.

If freespace propagation had been considered, then R_3 would have been selected as a relay node. This is because R_3 is within T_1 nominal radio coverage and R_3 is the farthest node (see Figure 1(a)). However, R_3 may not be the best option when considering the highway path loss model described by Equation (1), since T_1 and R_3 are moving in opposite directions and R_3 is near the border. Thus, the MFWAR mechanism adjusts T_1 radio coverage by means of β_s for vehicles belonging to set V_s, as is the case for R_3. Therefore, T_1 radio range is adjusted to the $\beta_s * R_n$ value and R_3 is not selected because it is out of range (see Figure 1(b)). On the other hand, R_1 and R_2 are vehicles that belong to the V_c set and MFWAR adjusts the radio range for these vehicles to $\beta_c * R_n$. Note that $\beta_c < 1$ in order to avoid selecting vehicles near the border which may lead to significant packet losses. Although R_1 and R_2 are within the adjusted $\beta_c * R_n$ radio range, R_2 is the vehicle most forward (see Figure 1(c)) and therefore it is chosen as a relay node (FN_{n+1})

Location Service

The FPBR protocol invokes the location service (LS) when a packet arrives for a vehicle with unknown geographic position. The LS is performed with broadcast packets. Therefore, in order to eliminate the broadcast storm problem, a suppression broadcast technique is included in FPBR. Additionally, a redundancy strategy is also introduced in order to improve the LS robustness. The proposed algorithm for the LS is detailed in the next subsections:

Destination Discovery

Three classes of nodes are considered in the destination discovery procedure: source nodes, relay nodes, and destination nodes. The source nodes are vehicles with data to send. The relay nodes are vehicles located along the dissemination path. This nodes relay destination discovery packets. The destination nodes are the vehicles towards which the data packets are sent.

If the geographic position of a destination node is unknown and a source node wants to send data packets to that vehicle, then a location request packet (LREQ) is broadcasted. This packet is used to find the geographic position of the destination node. Before broadcasting the LREQ, a transmitter (*i.e.*, the source node or a relay node) selects a set of vehicles as relay nodes through FPBR's selection algorithm (see Section 3.3.2). If the vehicle is the source node, then the set of relay nodes in the

LREQ packet must include at least one vehicle for each direction. The same applies for relay nodes located at intersections. On the other hand, relay nodes not located at intersections select vehicles located towards the dissemination direction before broadcasting the LREQ packet.

When the LREQ packet arrives at the destination node, it waits for twice the propagation time, τ, before sending a location reply packet (LREP). This waiting time is introduced in order to avoid possible packet collisions with LREQ packets that could arrive from different paths. Thus, after 2τ s the destination node generates the LREP containing its geographic position and speed. The LREP packet is disseminated towards the source node position using the same algorithm as the LREQ. When the LREP packet reaches the source, the data packet is forwarded towards the geographic position of the destination node by means of the MFWAR mechanism. In order to avoid collisions between LS packets and hello beacons, every node that hears a LS packet delays its beacon period by τ. Vehicles along the discovery messages (LREP and LREQ) path are aware of these packets since they are broadcast packets.

Redundancy Strategy

As previously mentioned, the aim of the MFWAR selection mechanism is to reduce the loss of discovery packets by considering the hard constraints imposed by the radio channel. Nevertheless, sometimes the discovery packets may not reach the intended primary relay node because of larger than expected path losses or inaccurate position predictions made by the selection algorithm (see Figure 2(a)). Therefore, if the discovery packet sent by the current forwarder node does not reach the primary relay node, then the FPBR algorithm will use a redundancy strategy that considers the use of a backup node, *BN*. The task of *BN* is to rebroadcast the discovery packet towards the primary relay node in order to continue with the discovery packet dissemination (see Figure 2(b)). As such, it is important to mention that *BN* does not modify the primary relay nodes table in the discovery packets. Note that the backup node must be located between the current forwarder node (named FN_n) and the primary relay node (named FN_{n+1}) in order to improve the reception probability. Besides choosing a backup node, an acknowledgment mechanism is needed to implement the redundancy strategy. However, in the IEEE 802.11a and IEEE 802.11p MAC layer there is no acknowledgment mechanism for broadcast packets. Therefore, a light acknowledgment mechanism is also included in FPBR. This mechanism and the backup node selection procedure are explained next.

$\bullet FN_n$ $\blacklozenge BN$ $\square FN_{n+1}$
(a)

$\bullet FN_n$ $\blacklozenge BN$ $\square FN_{n+1}$
(b)

Figure 2. Example of the redundancy strategy implemented in FPBR. (**a**) The discovery packet reaches the *BN* node but it does not reach the FN_{n+1} node; (**b**) The *BN* node rebroadcasts the discovery packet towards the primary relay node since the original discovery packet did not reach the FN_{n+1} node.

Light Acknowledgment Mechanism
If the primary relay node, FN_{n+1}, receives the discovery packet (see Figure 3(a)), then it will broadcast the discovery packet (with updated primary relay nodes entries) towards a new primary relay node (e.g., FN_{n+2}). As *BN* is located between the current forwarder, FN_n, and the primary relay node, FN_{n+1}, *BN* should be able to receive FN_{n+1} broadcast transmission (see Figure 3(b)). This feature is used in FBPR as an implicit acknowledgment procedure. Thus, if a node receives a discovery packet where its ID is listed as that belonging to a *BN*, then it will wait for 2τ s. If after this time *BN* does not detect the transmission of a broadcast packet from the FN_{n+1} node listed on the discovery packet, then it will rebroadcast the original discovery packet towards the FN_{n+1} node. It is important to note that as an implicit acknowledgment approach is being used in FPBR, additional control packets are not required. Therefore, the FPBR acknowledgment mechanism does not introduce additional overhead.

*FN_n *BN *FN_{n+1}
(a)

*FN_n *BN *FN_{n+1}
(b)

Figure 3. Example of the Light Acknowledgement Mechanism implemented in FPBR. (**a**) The primary relay node successfully receives the discovery packet; (**b**) The *BN* node successfully receives the implicit acknowledgment from FN_{n+1}.

Backup Nodes Selection

As previously mentioned, the *BN* nodes specific task is to rebroadcast the original discovery packet if the FN_{n+1} node does not perform its corresponding broadcast transmission. In FPBR, *BN* is selected in such way that it is located between FN_n and FN_{n+1}. Although choosing a *BN* right on the middle between the original transmitter and the chosen primary relay node may be a good starting point, such a simple selection mechanism does not consider the vehicles direction and the consequent path loss increase/decrease. Thus, FPBR implements the MFWAR mechanism to select the backup nodes as well. However, for the selection of the *BN* node the nominal radio range is assumed to be equal to the distance between the FN_n and FN_{n+1}. This distance will be referred as d (FN_n, FN_{n+1}).

For the *BN* node selection through MFWAR, vehicles are grouped in three sets based on their relative direction to the original forwarder node: V_{Ra}, V_{Rc} and V_{Rs}. Additionally, three dynamic factors β_{Ra}, β_{Rc} and β_{Rs} are considered. These factors are used to adjust d (FN_n, FN_{n+1}) by means of MFWAR. If the MFWAR mechanism cannot find a *BN* within the required range, then the current forwarder node is chosen as the *BN* node. As previously mentioned, after the selection of a *BN* node its ID is included in the corresponding table in the discovery packet.

An example of the BN node selection through the MFWAR mechanism is schematized in Figure 4. Assume that the distance from the original transmitter, FN_n, to the primary relay node, FN_{n+1}, is $d(FN_n,FN_{n+1})$. As both backup nodes candidates in Figure 4 are moving in the same direction as FN_n, β_{Rc} is used to adjust $d(FN_n, FN_{n+1})$. In Figure 4 the BC_2 node is nearer to the border of the adjusted distance than the BC_1 node. Therefore, BC_2 is chosen as BN for this example.

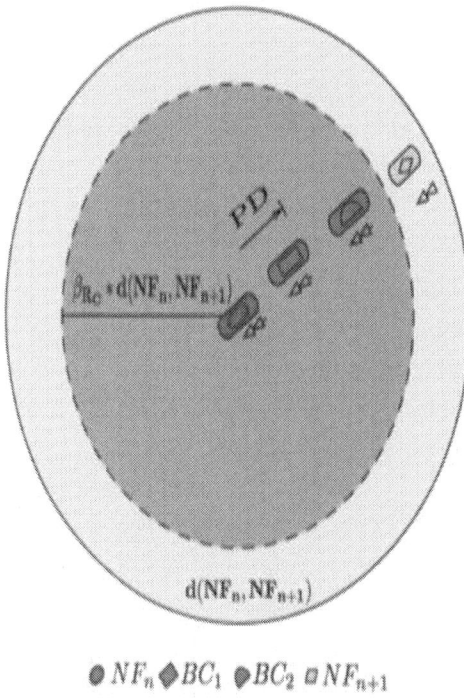

$$\bullet\, NF_n \quad \blacklozenge\, BC_1 \quad \bullet\, BC_2 \quad \square\, NF_{n+1}$$

Figure 4. Backup node (BN) selection through the MFWAR mechanism. The BC_2 node is chosen as BN because it is the nearest to the border of the adjusted distance $\beta_{Rc} * d(FN_n,FN_{n+1})$.

Data Dissemination

After the source node receives the destination geographic position, the data is sent towards the destination using the MFWAR mechanism. It is important to note that as the destination geographic position is included in the data packet header, every relay node can dynamically decide the next hop for the data packet. Furthermore, as every node has a digital map, the anchor-based approach [47,48] can be used where an intersection is found. In FPBR it is assumed that the IEEE 802.11a or the IEEE 802.11p RTS/CTS mechanism is disabled for data dissemination, this assumed to decrease the delay of the data packet and the overhead. Furthermore, implementing the RTS/CTS mechanism requires the transmission of several control packets before sending the actual data. A drawback of this strategy is that one or more of the RTS/CTS control packets could be lost because of the constraints imposed by the highway channel. This may lead to significant performance drops because at least 4 packets (RTS + CTS + DATA + ACK) must be sent to perform a single data packet transmission.

FPBR State Machine and Flow Chart

In order to explain the interaction among the different FPBR modules (*i.e.*, Hello_Mechanism, Location Service and Data Forwarding) a flow chart for each one of these modules is depicted in Figure 5. The Figure 5(a) presents the Hello Mechanism flowchart (named Hello Proc module). This module is in charge of sending the hello messages. The waiting period, T1, used in this module is calculated in accordance with the fixed beacon rate mentioned in Section 3.2. The Location Service flowchart is shown in Figure 5 (b) (named Loc_Serv). This module is in charge of performing the procedures described in Section 3.4. Lastly, the flowchart for the data forwarding mechanism explained in Section 3.5 is introduced in Figure 5 (c) (named Data_Fwd).

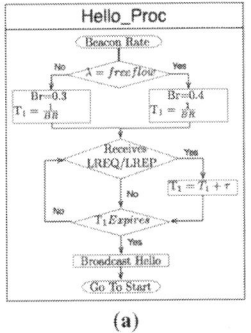 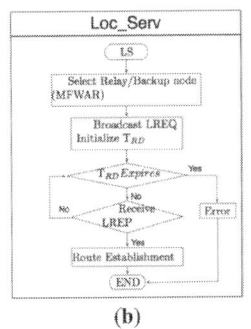 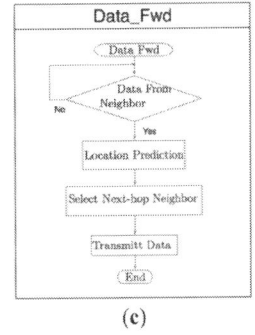

(a) (b) (c)

Figure 5. Flow charts for the main processes of the FPBR protocol: (**a**) The Hello mechanism; (**b**) the location service and (**c**) the data dissemination service.

The interaction between the different FPBR modules is controlled by means of a state machine (SM), as depicted in Figure 6. Every transition of the SM is in the format condition/procedure. A specific module from Figure 6 is launched when the corresponding condition of the state machine is fulfilled. The Hello timer condition is satisfied when the T1 waiting period ends, then the Hello proc is invoked. If the network layer receives a data packet from the application layer, then the App_pkt condition is fulfilled and the Loc_Serv module is invoked. Finally, if a node receives a data packet sent by a neighbor node, the Nb App_pkt condition is satisfied and the Fwd_Data module is invoked.

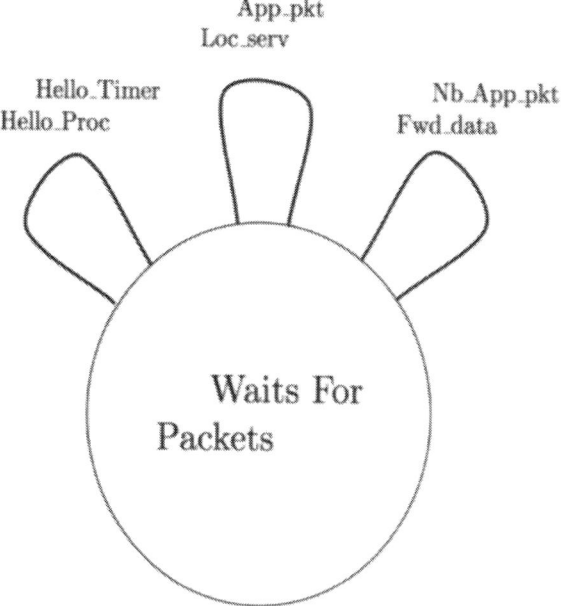

Figure 6. State machine of the interaction between the different modules of FPBR protocol.

PERFORMANCE EVALUATION

This section introduces several performance metrics obtained when using the FPBR protocol. These performance metrics are compared with those obtained when using the DDOR protocol for benchmark purposes. Particularly, metrics such as packet delivery ratio, delay per hop, overhead,

MAC overhead, end-to-end delay and number of hops were obtained for both protocols by means of a simulation setup programmed in the OPNET Modeler simulator. Two different propagation models were considered for the simulation setup: freespace propagation and the highway propagation model introduced in [37] for VANETs. For the simulation setup the IEEE 802.11a OPNET model was modified to resemble the IEEE 802.11p PHY/MAC layers as described below.

Benchmark Routing Protocol

The DDOR protocol has been used in this work as a benchmark for the performance of FPBR. The DDOR protocol was chosen because, as mentioned in the introduction, it was explicitly developed for the deployment of VANETs in highway scenarios. As such, in [39] it was shown that DDOR outperforms AODV, DSR, and GPSR. Furthermore, DDOR includes features specifically designed to deal with some of the impairments imposed by highway scenarios, which are not considered by other V2V protocols such as GPSR-L, LORA-CBF and CAR.

The DDOR protocol includes a discovery service coupled with a packet retransmission strategy. As previously mentioned, including these strategies in a VANET routing protocol for highways is very important, since the wireless channel constraints imposed by these scenarios may lead to significant packet losses. The DDOR protocol proposes enabling the IEEE 802.11 RTS/CTS mechanism for its location service and for the data dissemination stages. Thus, DDOR implements a unicast strategy for its location service. These unicast packets are sent in all directions in order to discover the geographic position of the destination node. Additionally, DDOR proposes an adaptive beacon mechanism (AB) based on the traveled distance of the vehicle. The AB mechanism goal is to reduce the overhead caused by the hello messages. Shortly, the AB mechanism sends a beacon message every time that a vehicle crosses a predefined milestone. Finally, DDOR includes a location prediction algorithm based on the vehicle relative velocity and the size of the transmitted packet. This location prediction algorithm is used before sending any packet in order to select the next-hop vehicle.

In order to compare FPBR and DDOR under the same scenario conditions both protocols were implemented in the OPNET Modeler simulator [69]. The FPBR implementation follows the state machine and flowchart introduced in Section 3.6. The DDOR protocol was implemented following the guidelines provided in [39].

Evaluation Setup

IEEE 802.11a OPNET Model PHY/MAC Adaptation to IEEE 802.11p PHY/MAC Parameters

As previously mentioned, the simulation testbed for FPBR and DDOR was programmed in the OPNET Modeler simulator. This is a well recognized simulation suite widely used in the industry and academia for the evaluation of communication networks, [35,70,71]. However, by the time this research work was completed, a specific model for the IEEE 802.11p standard was not included in the OPNET Modeler (v.16.0). Nevertheless, because of the relevance of the IEEE 802.11p standard within the context of V2V communications, the simulation setup used an adaptation of the IEEE 802.11a model such that all PHY/MAC settings and parameters correspond to those found in the IEEE 802.11p standard. This approach was previously used in [35] for the evaluation of AODV and DSR in VANETs equipped with IEEE 802.11p transceivers.

As mentioned in the introduction, the MAC layer in IEEE 802.11p adopted the IEEE 802.11e EDCA mechanism. This mechanism defines four different classes of traffic with different distributed coordination function (DCF)-CSMA/CA parameters, [20,35,36]. In this paper IEEE 802.11p DCF parameters corresponding to best effort traffic over service channels (see [36]) were used to modify the IEEE 802.11a OPNET model. Therefore, the minimum and maximum contention windows sizes, and the time slot length were adjusted to IEEE 802.11p best effort traffic values. Similarly, the DIFS value was replaced by the corresponding AIFS value. Regarding the PHY layer adaptation, the bandwidth and operating frequency of the IEEE 802.11a OPNET model were adjusted to 10 MHz and 5.880 GHz respectively, as defined for IEEE 802.11p standard. Additionally, the data rates of the IEEE 802.11a OPNET model were adjusted to those defined by the IEEE 802.11p standard. For the rest of this paper the modified model will be referred as adapted IEEE 802.11a/p model.

Simulation Scenario

The performance of both protocols was evaluated considering a highway scenario with two lanes for cars travelling in one direction and other two lanes for cars travelling in the opposite direction. The length of each lane was set to 3 km and the width to 4 m. When a vehicle reaches the end of the road, it is reinserted in the lane with vehicles traveling in the opposite direction. The maximum allowed speed was set to 60 m/s. The radio channel propagation model introduced in [37] was implemented in the simulation setup. Both protocols were tested considering this model which, as previously mentioned, was specifically developed for V2V highway scenarios. Additionally, performance metrics for both protocols were obtained considering the freespace propagation model. The mobility pattern for each vehicle was generated following the intelligent driver model introduced in [66]. This is a popular model used to generate mobility patterns for highway and urban scenarios, (e.g., [50,72]). Furthermore, as the vehicles acceleration is an important parameter that could modify the network topology [21], both protocols were evaluated under various maximum vehicle acceleration rates ranging from $\alpha = 1.6$ m/s^2 to $\alpha = 5$ m/s^2. Free flow, medium, high and jam vehicle densities were considered in order to evaluate the effects of the vehicle density, λ, in the routing metrics. The specific value of λ for each density was taken from [67] as shown in Table 1. The adapted IEEE 802.11a/p model with data rate of 6 Mbps was considered for the vehicle transceiver. The performance metrics for each protocol were obtained considering four transmitters. At the beginning of every simulation trial, every transmitter is set to wait for 5 s before starting any transmission. This 5 s period was set to better allow the exchange of hello messages. When the waiting period concludes each transmitter chooses a random starting time and a random destination node. Once a vehicle starts transmitting, it generates packets for an equivalent user data rate of 2 Kbytes per second. This data rate is maintained during the entire simulation. A minimum of 100 trials were performed for each acceleration-density pair. Each trial lasted 300 s during which different performance metrics were recorded. The particular values of each variable used in the simulation are presented inTable 1.

Table 1. Specific values used for each variable in the simulation scenario.

Parameter	Value
Maximum Velocity	60 m/s
Maximum Acceleration	$[1.6–5.0]$ m/s^2
Vehicle Density	$[33,66,100,133]$ vehicles/km
Highway Length	3,000 m
Number of Lanes	4
Mobility Model	IDM
Simulation Time	300 s
Packet Size	2 Kbytes
Base Frequency	5.880 GHz
Data Rate	6 Mbps
Transmission Range	300 mts

Metrics

The FPBR and DDOR performance was evaluated using the following metrics:

- Packet delivery ratio (PDR). This metric is measured as the ratio between the number of data packets received by the destination and the number of data packets transmitted by the source.
- Delay per hop. This is the average time a data packet requires for a single hop transmission.
- Average end-to-end delay (EED). The EED is the average time for a data packet to reach its destination.
- Network overhead. This metric is measured in terms of the number of routing control packets transmitted per second normalized by the

vehicle density. This metric includes the number of beacons plus the discovery control packets.

- MAC Overhead. This metric is measured in terms of the number of control packets that the MAC layer transmits per second normalized by the vehicle density.
- Number of hops (NH). The NH is the average number of hops required for a data packet to reach its destination.

Results and Analysis

Two different radio propagation models were considered for the evaluation of DDOR and FPBR. Thus, the performance metrics are presented in two subsections, one for each propagation model. The results obtained when assuming freespace propagation are introduced in Subsection 4.4.1. Additionally, Subsection 4.4.2 presents the metrics obtained when considering the radio propagation model explicitly developed for highway scenarios introduced in [37].

Results Considering the Freespace Channel

As discussed in Section 3, the freespace propagation model does not hold for the radio channel conditions present in highway scenarios as measured for V2V communications [37]. However, it is an important scenario commonly used for evaluation purposes [55]. As no fading is introduced when considering the freespace channel, every transmitted packet will reach the transceiver nominal radio range. Thus, for this channel the collision probability for broadcast packets is higher than that observed when considering a fading channel. Therefore, the effect of the packets collisions on the routing strategy performance can be observed with more detail when assuming freespace propagation. The routing metrics obtained when increasing the maximum acceleration range, α, for two different densities, λ, are shown inFigures 7–9. It can be seen in Figure 7(a) that the packet delivery ratio (PDR) is above 98% for both protocols. Because of the transmitting radio range is deterministic for freespace propagation, any packet loss is caused by collisions or erroneous predictions for both protocols. Thus, as Figure 7(a) shows, both kinds of packet losses are adequately handled by the particular redundancy strategy implemented in DDOR and FPBR. Similar conclusions can be drawn for the number of hops (NH), the end to end delay (EED) and the average delay per hop as shown in Figure 7(b), Figure 8(a,b) respectively. In fact, it can be readily seen in Figures 7 and 8 that both protocols perform similarly well when considering the PDR, NH, EED, and delay per hop metrics.

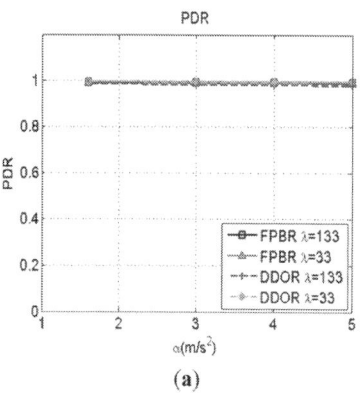 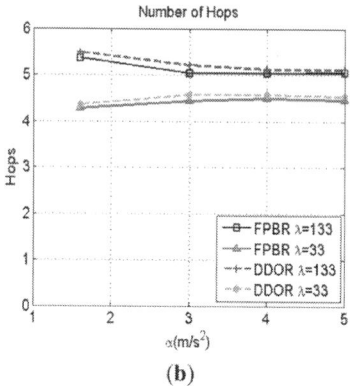

(a)   **(b)**

Figure 7. Packet delivery ratio (PDR) and number of hops (NH) obtained with FPBR and DDOR when increasing the maximum acceleration, α. Two different vehicle densities were considered for these plots: $\lambda = 133$ and $\lambda = 33$. (**a**) PDR *vs.* α; (**b**) Number of Hops *vs.* α.

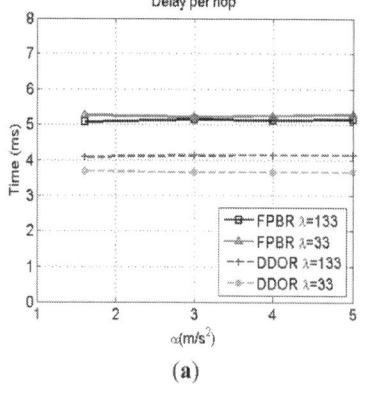 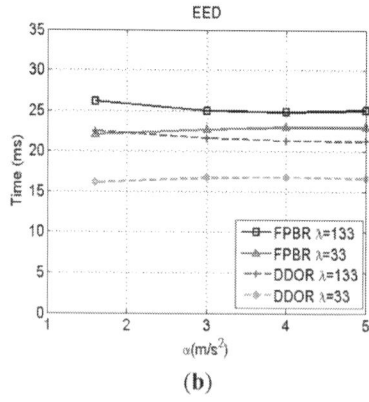

(a)   **(b)**

Figure 8. Delay per hop and end-to-end delay (EED) obtained with FPBR and DDOR when increasing the maximum acceleration, α. Two different vehicle densities were considered for these plots: $\lambda = 133$ and $\lambda = 33$. (**a**) Delay per hop *vs.* α. (**b**) EED *vs.* α.

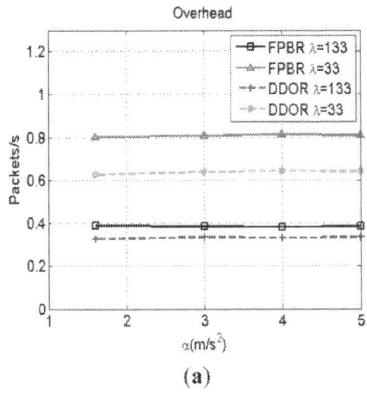 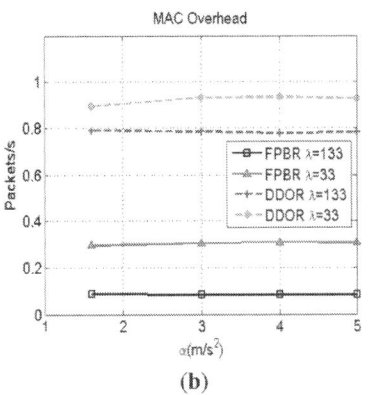

Figure 9. Network and MAC overhead obtained with FPBR and DDOR when increasing the maximum acceleration, α. Two different vehicle densities were considered for these plots: $\lambda = 133$ and $\lambda = 33$. (a) Network overhead vs.α. (b) MAC overhead vs. α.

The network and the MAC overhead metrics are presented in Figure 9. According with Figure 9(a), DDOR generates fewer control packets than FPBR at the network layer. This behavior is observed because DDOR implements its redundancy strategy at the MAC layer by using the RTS/CTS mechanism. On the other hand, Figure 9(b) shows that FPBR yields better MAC overhead metrics than DDOR. This behavior is observed because the RTS/CTS mechanism used in DDOR needs a minimum of three additional packets to transmit a single data packet.

V2V Channel Propagation Model
This subsection presents the metrics obtained when considering the radio propagation model explicitly developed for highway scenarios introduced in [37]. Remember that for this channel a transmitted packet may not reach the transceiver nominal radio range because of the path loss and the shadowing (see Section 3.1). Therefore, for this case a packet may not reach its intended destination because of collisions or adverse channel conditions. The Figure 10 presents the PDR obtained with FPBR and DDOR when increasing the maximum vehicle acceleration, α. Four different vehicle densities, λ, were considered for this figure. Note in Figure 10 that DDOR provides a much lower PDR than that observed

when considering freespace propagation. In contrast, the PDR obtained when using FPBR shows a similar performance to that observed in Figure 7(a). This performance drop for DDOR is caused by the underestimation of the V2V highway channel effects when selecting the next forwarder node. As explained in Section 3, FPBR considers the specific constraints imposed by the V2V highway channel when selecting the next forwarder node. Thus, when considering the V2V highway channel FPBR does not show a significant performance drop, compared to the performance achieved when considering the freespace channel. Additionally, note that the PDR is not significantly affected by the acceleration rate when using FPBR. This means that the backup mechanism implemented in FPBR is able to cope with packet collisions and erroneous predictions made by the LP mechanism.

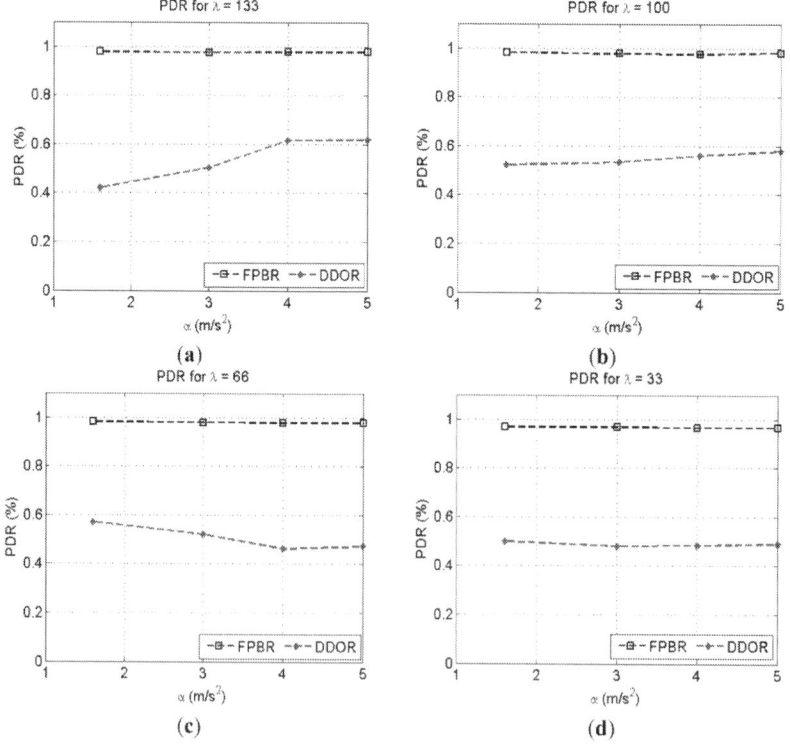

Figure 10. Packet Delivery Ratio (PDR) obtained with FPBR and DDOR when increasing the maximum acceleration, α. Four different vehicle densities, λ, were considered for these plots. Each plot shows the PDR *vs.* α for: **(a)** $\lambda = 133$; **(b)** $\lambda = 100$; **(c)** $\lambda = 66$ and **(d)** $\lambda = 33$.

The Figure 11 shows the PDR obtained by FPBR and DDOR for different source-destination distances. The plots in Figure 11 were obtained by setting $\alpha = 4$ with $\lambda = 66$ and $\lambda = 133$ Importantly, note in this figure how the PDR obtained with DDOR exhibits a decreasing behavior as the source-destination distance increases. This is caused by erroneous next-hop selections made by DDOR (remember that for this protocol the channel effects are not considered when performing the next-hop selections). In contrast, FPBR provides a significantly higher PDR when compared to DDOR. The FPBR protocol exhibits this behavior because of its backup and next-hop selection mechanisms enable the delivery of packets traveling longer source-destination distances.

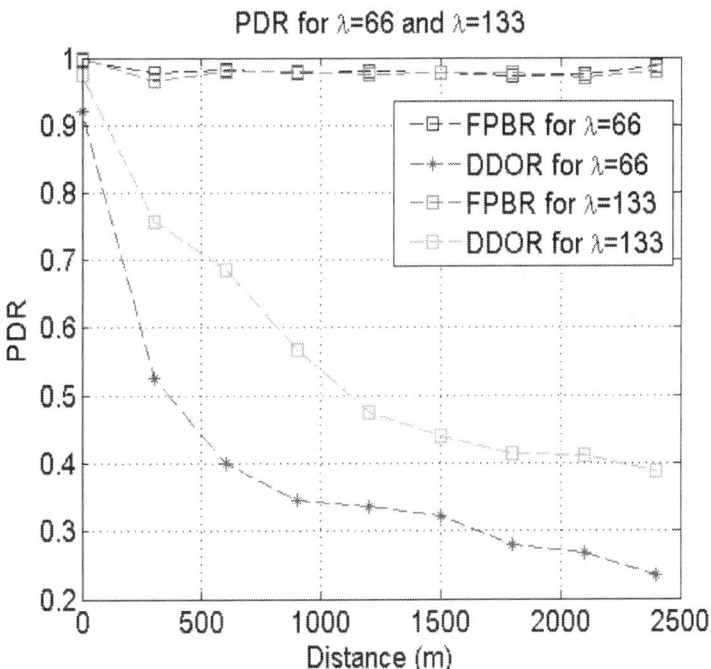

Figure 11. Packet Delivery Ratio (PDR) obtained with FPBR and DDOR when increasing the source-destination distance. The maximum vehicle acceleration was set to $\alpha = 4$. Two different vehicle densities, $\lambda = 66$ and $\lambda = 133$, were considered for these figures.

Figure 12 illustrates the effects of increasing the maximum acceleration, α, on the average number of hops per packet for different vehicle densities, λ.

It can be seen in this figure that DDOR has a lower hop count. Although at first this may appear to be an advantage, the reason for the lower hop count in DDOR is the increase in the packet drop probability observed when the source-destination distance is incremented (see Figure 11). On the other hand, the hop count of FPBR is larger because it enables the delivery of packets traveling longer source-destination distances (see Figure 11).

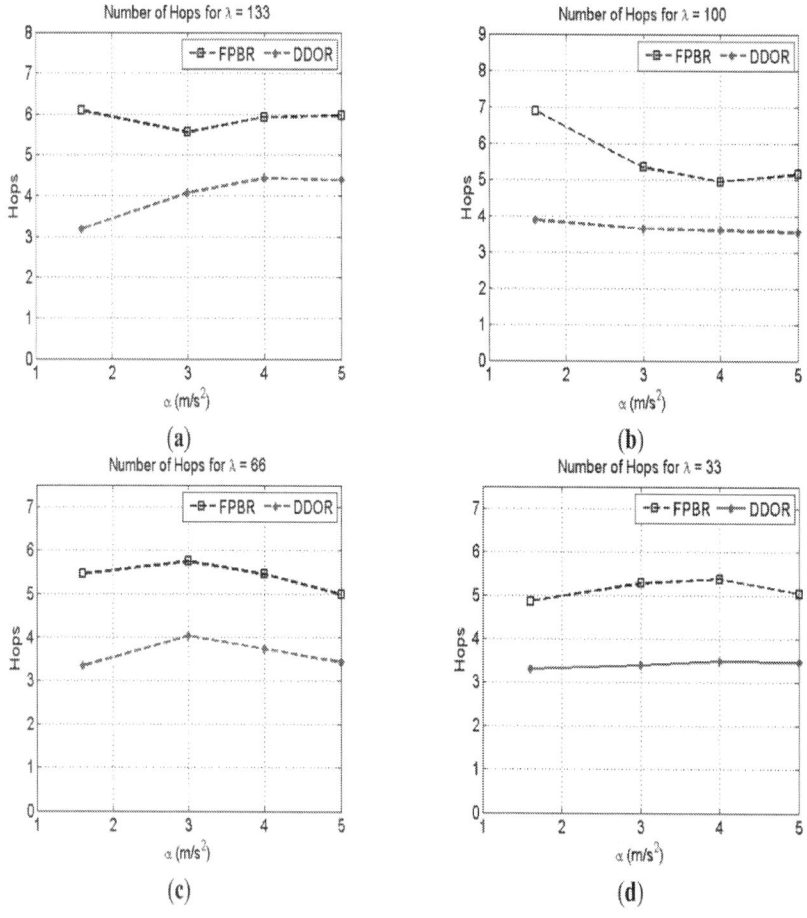

Figure 12. Average number of hops obtained with FPBR and DDOR when increasing the maximum acceleration, α. Four different vehicle densities, λ, were considered for these plots. Each plot shows the number of hops *vs.* α for: (**a**) $\lambda = 133$; (**b**) $\lambda = 100$; (**c**) $\lambda = 66$ and (**d**) $\lambda = 33$.

Figure 13 shows the effects of increasing the maximum acceleration, α, on the average delay per hop metric for different vehicle densities, λ. It can be seen that FPBR achieves a lower delay per hop than DDOR. This means that the next hop selection mechanism of FPBR made fewer erroneous selections than those made by the selection algorithm of DDOR. Furthermore, as previously mentioned DDOR's redundancy strategy needs at least 3 control packets while the redundancy strategy of FPBR implements an implicit acknowledgement mechanism. Therefore, when an erroneous selection is made, the redundancy strategy of DDOR generates higher medium access contention than that generated by FPBR's redundancy strategy.

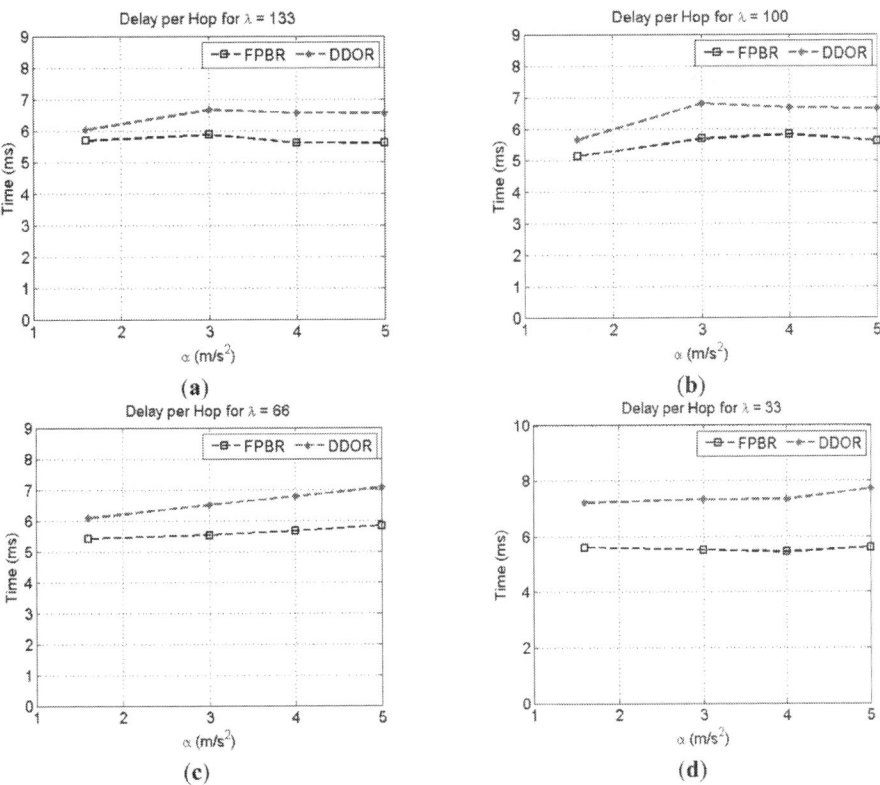

Figure 13. Average delay per hop obtained with FPBR and DDOR when increasing the maximum acceleration, α. Four different vehicle densities, λ, were considered for these plots. Each plot shows the delay per hop *vs.* α for: (**a**) $\lambda = 133$; (**b**) $\lambda = 100$; (**c**) $\lambda = 66$ and (**d**) $\lambda = 33$.

Figure 14 illustrates the EED metric behavior when increasing the maximum acceleration, α, for different vehicle densities, λ. It can be seen in this figure that when using DDOR and increase in α leads to an increase in the EED metric for $\lambda = 133$ and $\lambda = 100$ (*i.e.*, medium and jam vehicular densities). In contrast, when using FPBR the changes in the acceleration rate do not lead to an increase in the EED for these vehicular densities. The FPBR protocol shows this behavior because of its particular next-hop selection algorithm, described in Section 3.3.2. Thus, it can be inferred that by using the dynamic factors β_a, β_c and β_s the probability of performing an erroneous next hop selection is decreased. Note that for the free flow vehicle density, $\lambda = 33$, the change in the acceleration, α, does not seem to have a significant effect in the EED metric. This behavior is observed because the location prediction mechanism of both protocols is more accurate for this vehicle density.

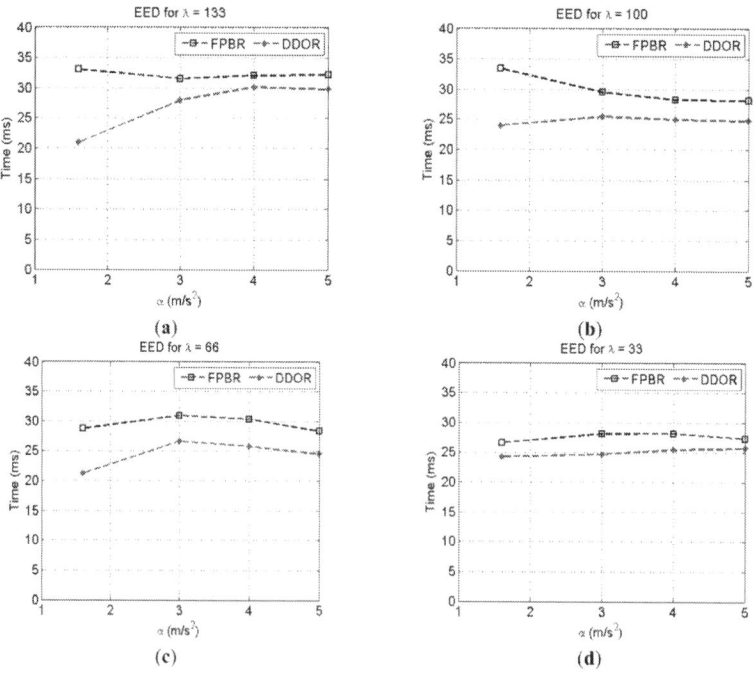

Figure 14. Average end-to-end delay (EED) obtained with FPBR and DDOR when increasing the maximum acceleration, α. Four different vehicle densities, λ, were considered for these plots. Each plot shows the EED *vs.* α for: **(a)** $\lambda = 133$; **(b)** $\lambda = 100$; **(c)** $\lambda = 66$ and **(d)** $\lambda = 33$.

Figure 15 shows the effects of increasing the maximum acceleration, α, on network overhead metric for different vehicle densities, λ. As previously mentioned, the redundancy strategy of DDOR is performed at the MAC layer. Thus, the overhead introduced by DDOR at the network layer mainly consists of hello messages. Consequently, the network overhead for DDOR is lower than that introduced by FPBR (seeFigure 15). Note that for $\lambda = 33$ both protocols show a higher network overhead (see Figure 15(d)). This is caused by the particular adaptive beacon rate mechanisms implemented in both protocols (see Sections 3.2 and 4.1).

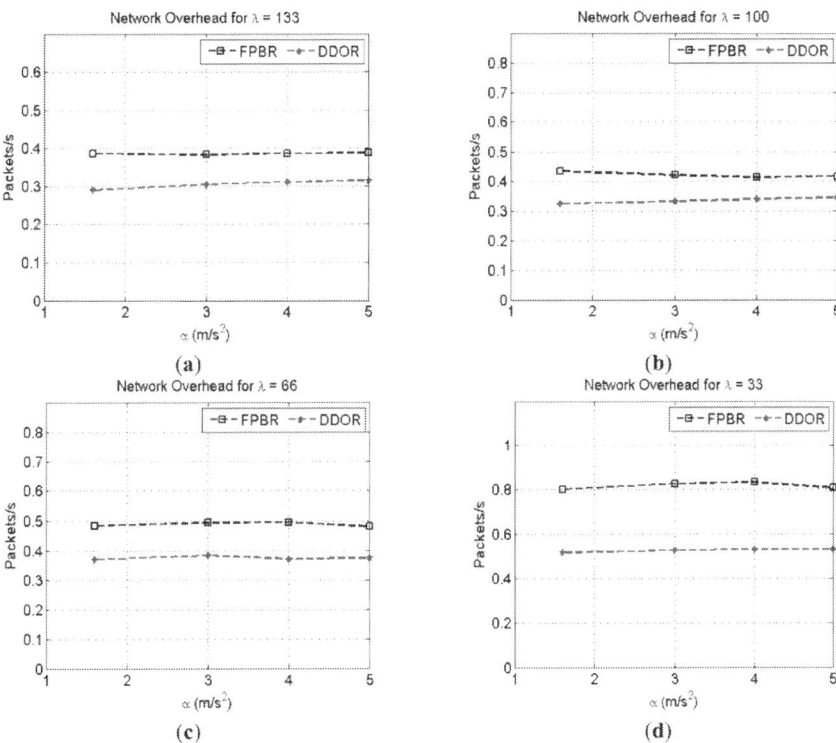

Figure 15. Normalized network overhead obtained with FPBR and DDOR when increasing the maximum acceleration, α. Four different vehicle densities, λ, were considered for these plots. Each plot shows the network overhead *vs.* α for: (**a**) $\lambda = 133$; (**b**)$\lambda = 100$; (**c**) $\lambda = 66$ and (**d**) $\lambda = 33$.

Lastly, the Figure 16 shows the MAC overhead obtained when increasing the maximum acceleration, α, for different vehicle densities, λ. As mentioned in Section 4.1, the RTS/CTS mechanism used in DDOR needs a minimum of three additional packets to transmit a single data packet. Consequently, this mechanism significantly increases the overhead introduced at the MAC layer. In contrast, FPBR provides a lower MAC overhead. This is because FPBR implements an implicit acknowledgement mechanism in the redundancy strategy for the location service, as explained in Section 3.4. Additionally, FPBR does not enable the RTS/CTS mechanism for the transmission of data packets. Thus, FPBR mitigates the effects of the erroneous selections with fewer packets than those needed by DDOR.

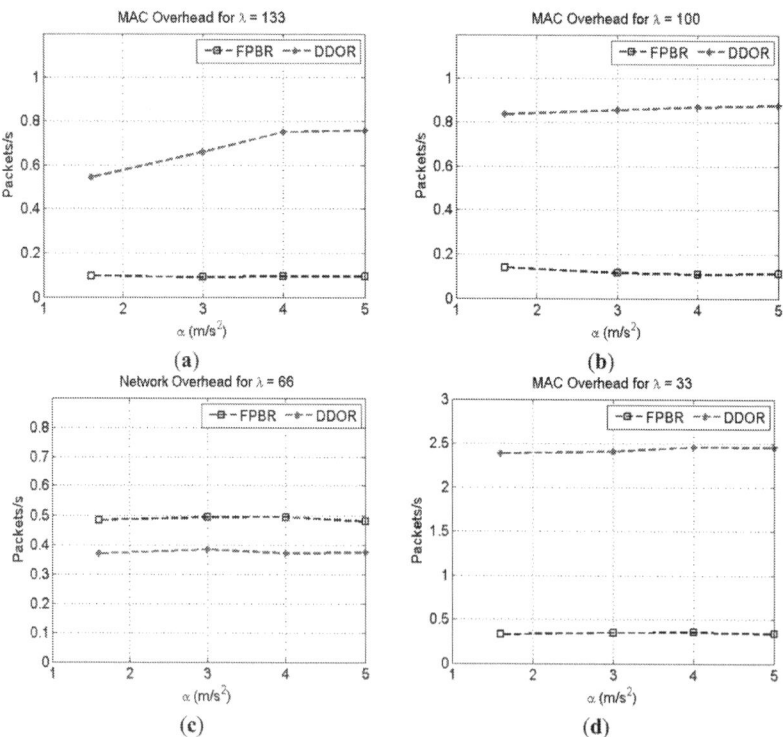

Figure 16. Normalized MAC overhead obtained with FPBR and DDOR when increasing the maximum acceleration, α. Four different vehicle densities, λ, were considered for these plots. Each plot shows the MAC overhead *vs.* α for: (**a**) $\lambda = 133$; (**b**) $\lambda = 100$; (**c**) $\lambda = 66$ and (**d**) $\lambda = 33$.

CONCLUSIONS AND FUTURE WORK

In this paper a new routing protocol for VANETs called Freestanding Position-Based Routing (FPBR) protocol was introduced. The design of FPBR was aimed to the deployment of VANET's in highway scenarios. As such, the performance of FPBR was compared with the performance shown by the DDOR protocol, which is one of the few protocols explicitly developed for V2V communications in highway scenarios. When considering freespace propagation conditions, FPBR delivers performance metrics similar to those obtained when using the leading DDOR protocol. Nevertheless, when considering the V2V highway channel, the performance metrics obtained with both protocols for this scenario show that FPBR outperforms DDOR. The reason for this is because FPBR includes several mechanisms which help to overcome packet losses caused by the harsh conditions imposed by the highway radio channel. Particularly, FPBR's next hop selection and the implicit acknowledgment mechanisms proved to be vital when considering the highway channel, since these routines enabled the selection of suitable primary relay and backup nodes. Consequently, these mechanisms allowed FPBR to yield a high packet delivery ratio. Furthermore, the simulation results show that neither the redundancy strategy nor the location service of FPBR adds a significant overhead, even though the location service uses broadcast packets. Therefore, with the results presented in Section 4.4 it has been shown that in order to improve the performance of the routing protocols for highway scenarios, the constraints imposed by the radio channel must be considered when designing the routing strategy. In that sense, FPBR is a suitable protocol for VANETs deployment in highway scenarios, as it offers several advantages compared to traditional VANETs protocols (e.g., AODV and GPSR-L) and even outperforms the leading DDOR protocol for this kind of scenarios. In this paper the β parameters have been chosen based on overall performance for different densities. The performance of FPBR could be improved if the β parameters are customized for different network conditions. Thus, future work includes developing an adaptive method to set the βparameters values based on current network conditions. In order to adapt FPBR for city scenarios, future work will include obtaining specific β parameters for urban scenarios. Regarding the IEEE 802.11p standard, FPBR was evaluated considering one traffic class, delivering satisfactory performance metrics. Similarly, a single traffic class was considered in [35] for the performance analysis of AODV and DSR when using IEEE 802.11p transceivers (as previously mentioned the IEEE 802.11p implementation used in [35] was based in the IEEE 802.11a OPNET model). Therefore, future work will include obtaining performance metrics for FPBR, DDOR, AODV and DSR when considering concurrent traffic flows with different IEEE 802.11p traffic classes.

This research work was possible thanks to scholarship number 167835 granted by the National Council of Science and Technology (CONACYT, Mexico). This work was partially supported by CONACYT Basic Science Grant No. 169333. Authors would like to thank David Covarrubias and Victor Rangel for the help provided for the realization of this work.

REFERENCES

1. *Statistics*; U.S. Department of Transportation, U.S. Government Printing Office: Washington, DC, USA, 2010.
2. Popescu-Zeletin, R.; Ilja Radusch, M.A.R. *Vehicular-2-X Communication: State-Of-The-Art and Research in Mobile Vehicular Ad Hoc Networks*; Springer: Berlin, Germany, 2010.
3. IEEE. Intelligent Transport Systems Society. Available online: http://sites.ieee.org/itss/ (accessed on 15 September 2012).
4. ITS-Europe. Available online: http://ec.europa.eu/transport/its (accessed on 8 October 2012).
5. Car2Car Communication Consortium. Avaliable online: http://www.car-to-car.org/index.php?id=46 (accessed on 8 October 2012).
6. Chu, Y.C.; Huang, N.F. An efficient traffic information forwarding solution for vehicle safety communications on highways. *IEEE Trans. Intell. Transp. Syst.* **2012**, *13*, 1524–9050.
7. Benslimane, A.; Barghi, S.; Assi, C. Fast track article: An efficient routing protocol for connecting vehicular networks to the Internet. *Pervasive Mob. Comput.* **2011**, *7*, 98–113.
8. Tonguz, O.K.; Boban, M. Multiplayer games over vehicular *ad hoc* networks: A new application.*Ad Hoc Netw.* **2010**, *8*, 531–543.
9. Isento, J.; Dias, J.; Neves, J.; Soares, V.; Rodrigues, J.; Nogueira, A.; Salvador, P. FTP@VDTN: A File Transfer Application for Vehicular Delay-Tolerant Networks. Proceedings of the 2011 International Conference on Computer as a Tool (EUROCON), Lisbon, Portugal, 27–29 April 2011; pp. 1–4.

10. Milanes, V.; Villagra, J.; Godoy, J.; Simo, J.; Perez, J.; Onieva, E. An intelligent V2I-based traffic management system. *IEEE Trans. Intell. Transp. Syst.* **2012**, *13*, 49–58.

11. Festag, A.; Noecker, G.; Strassberger, M.; Lubke, A.; Bochow, B.; Torrent-Moreno, M.; Schnaufer, S.; Eigner, R.; Catrinescu, C.; Kunisch, J. NoW—Network on Wheels: Project Objectives, Technology and Achievements. Proceedings of the 6th International Workshop on Intelligent Transportation, Hamburg, Germany, 24–25 March 2009.

12. Hartenstein, H.; Bochow, B.; Ebner, A.; Lott, M.; Radimirsch, M.; Vollmer, D. Position-Aware *Ad Hoc* Wireless Networks for Inter-Vehicle Communications: The Fleetnet Project. Proceedings of the 2nd ACM International Symposium on Mobile Ad Hoc Networking, Computing, Long Beach, CA, USA, 4– 5 October 2001; pp. 259–262.

13. Cooperative-Vehicle-Infrastructure-Systems. Available online: http://www.cvisproject.org(accessed on 15 September 2012).

14. Safespot. Available online: http://www.safespot-eu.org (accessed on 15 September 2012).

15. Operative Systems for Intelligent Road Safety, C. Available online: http://www.coopers-ip.eu(accessed on 15 September 2012).

16. eCoMove. Available online: http://www.ecomove-project.eu (accessed on 15 September 2012).

17. Barr, R. An Efficient, Unifying Approach to Simulation Using Virtual Machines. Ph.D. Thesis, Cornell University, Ithaca, NY, USA, 2004.

18. SOUNDBITES: Vehicle-To-Vehicle Communications. Available online: http://media.ford.com/article_display.cfm?article_id=33971 (accessed on 15 September 2012).

19. Kosch, T.; Kulp, I.; Bechler, M.; Strassberger, M.; Weyl, B.; Lasowski, R. Communication architecture for cooperative systems in Europe. *IEEE Commun. Mag.* **2009**, *47*, 116–125.

20. IEEE Standard for Information Technology. *Part 11: Wireless LAN Medium Access Control (MAC) and Physical Layer (PHY) Specifications: Amendment 6: Wireless Access in Vehicular Environments*; IEEE Standard No. 802.11-2007. IEEE: New York, NY, USA, 2007.

21. Abdel Hafeez, K.; Zhao, L.; Liao, Z.; Ma, B. Clustering and OFDMA-based MAC protocol (COMAC) for vehicular *ad hoc* networks. *EURASIP J. Wireless Commun. Netw.* **2011**, *2011*, 117.

22. Abumansoor, O.; Boukerche, A. A secure cooperative approach for nonline-of-sight location verification in VANET. *IEEE Trans. Veh. Tech.* **2012**, *61*, 275–285.

23. Lu, R.; Li, X.; Luan, T.; Liang, X.; Shen, X. Pseudonym changing at social spots: An effective strategy for location privacy in VANETs. *IEEE Trans. Veh. Tech.* **2012**, *61*, 86–96.

24. Bernsen, J.; Manivannan, D. Unicast routing protocols for vehicular ad hoc networks: A critical comparison and classification. *Pervasive Mobile Comput.* **2009**, *5*, 1–18.

25. Taysi, Z.; Yavuz, A. Routing protocols for GeoNet: A survey. *IEEE Trans. Intell. Transp. Syst.* **2012**, *13*, 939–954.

26. Fonseca, A.; Vazao, T. Applicability of position-based routing for VANET in highways and urban environment. *J. Netw. Comput. Appl.* **2012**.

27. Fernndez-Carams, T.; Gonzlez-Lpez, M.; Castedo, L. Mobile WiMAX for vehicular applications: Performance evaluation and comparison against IEEE 802.11p/a. *Comput. Netw.* **2011**, *55*, 3784–3795.

28. Mizutani, K.; Kohno, R. Inter-vehicle spread spectrum communication and ranging system with concatenated EOE sequence. *IEEE Trans. Intell. Transp. Syst.* **2001**, *2*, 180–191.

29. Inoue, T.; Nakata, H.; Itami, M.; Itoh, K. An Analysis of Incident Information Transmission Performance Using an IVC System That Assigns PN Codes to the Locations on the Road. Proceedings of the 2004 IEEE Intelligent Vehicles Symposium, Parma, Italy, 14– 17 June 2004; pp. 115–120.

30. Nagaosa, T.; Hasegawa, T. An autonomous Distributed Inter-Vehicle Communication Network Using Multicode Sense CDMA. Proceedings of the 5th International Symposium on Spread Spectrum Techniques and Applications, Sun City, South Africa, 2– 4 September 1998; Volume 3. pp. 738–742.

31. Sawant, H.; Tan, J.; Yang, Q.; Wang, Q. Using Bluetooth and Sensor Networks for Intelligent Transportation Systems. Proceeding of the 7th International IEEE Conference on Intelligent Transportation Systems, Washington, DC, USA, 4– 6 October 2004; pp. 767–772.

32. Pasolini, G.; Verdone, R. Bluetooth for ITS? Proceeding of the 5th International Symposium on Wireless Personal Multimedia Communications, Honolulu, HI, USA, 27–30 October 2002; Volume 1. pp. 315–319.

33. Sugiura, A.; Dermawan, C. In traffic jam IVC-RVC system for ITS using Bluetooth. *IEEE Trans. Intell. Transp. Syst.* **2005**, *6*, 302–313.

34. Fernandez, J.; Borries, K.; Cheng, L.; Kumar, B.; Stancil, D.; Bai, F. Performance of the 802.11p physical layer in vehicle-to-vehicle environments. *IEEE Trans. Veh. Tech.* **2012**, *61*, 3–14.

35. Iqbal, M.; Wang, F.; Xu, X.; Eljack, S.; Mohammad, A. Reactive routing evaluation using modified 802.11a with realistic vehicular mobility. *Ann. Telecommun.* **2011**, *66*, 643–656.

36. Mišić, J.; Badawy, G.; Mišić, V. Performance characterization for IEEE 802.11p network with single channel devices. *IEEE Trans. Veh. Tech.* **2011**, *60*, 1775–1787.

37. Karedal, J.; Czink, N.; Paier, A.; Tufvesson, F.; Molisch, A. Path loss modeling for vehicle-to- vehicle communications. *IEEE Trans. Veh. Tech.* **2011**, *60*, 323–328.

38. Baccelli, E.; Jacquet, P.; Mans, B.; Rodolakis, G. Highway vehicular delay tolerant networks: Information propagation speed properties. *IEEE Trans. Inform. Theor.* **2012**, *58*, 17431756.

39. Sahu, P.; Wu, E.; Sahoo, J.; Gerla, M. DDOR: Destination discovery oriented routing in highway/freeway VANET+. *Telecommun. Syst. 2010*.

40. Rawashdeh, Z.; Mahmud, S. A novel algorithm to form stable clusters in vehicular *ad hoc* networks on highways. *EURASIP J. Wireless Commun. Netw.* **2012**, *2012*, 15.

41. Vinel, A. 3GPP LTE *versus* IEEE 802.11p/WAVE: Which technology is able to support cooperative vehicular safety applications? *IEEE Wireless Commun. Lett.* **2012**, *1*, 125–128.

42. Yan, Z.; Jiang, H.; Shen, Z.; Chang, Y.; Huang, L. k-Connectivity analysis of one-dimensional linear VANETs. *IEEE Trans. Veh. Tech.* **2012**, *61*, 426–433.

43. Abedi, O.; Fathy, M.; Taghiloo, J. Enhancing AODV Routing Protocol Using Mobility Parameters in VANET. Proceedings of the 2008. IEEE/ACS International Conference on Computer Systems and Applications, Doha, Qatar, 31 March–4 April 2008; pp. 229–235.

44. Johnson, D.B.; Maltz, D.A.; Broch, J. DSR: The dynamic source routing protocol for multi-hop wireless *ad hoc* networks. In *Ad Hoc Networking*; Addison-Wesley Professional: White Plains, NY, USA, 2001; pp. 139–172.

45. Karp, B.; Kung, H.T. GPSR: Greedy Perimeter Stateless Routing for Wireless Networks. Proceedings of the 6th Annual International Conference on Mobile Computing and Networking, Boston, MA, USA, 6–11 August 2000; pp. 243–254.

46. Chaurasia, B.; Tomar, R.; Verma, S.; Tomar, G. Suitability of MANET Routing Protocols for Vehicular *Ad Hoc* Networks. Proceedings of the 2012 International Conference on Communication Systems and Network Technologies (CSNT), Rajkot, India, 11– 13 May 2012; pp. 334–338.

47. Naumov, V.; Gross, T. Connectivity-Aware Routing (CAR) in Vehicular *Ad-Hoc* Networks. Proceedings of the 26th IEEE International Conference on Computer Communications (INFOCOM), Anchorage, AK, USA, 6– 12 May 2007.

48. Jerbi, M.; Senouci, S.M.; Rasheed, T.; Ghamri-Doudane, Y. Towards efficient geographic routing in urban vehicular networks. *IEEE Trans. Veh. Tech.* **2009**, *58*, 5048–5059.

49. Rao, S.; Pai, M.; Boussedjra, M.; Mouzna, J. GPSR-L: Greedy Perimeter Stateless Routing with Lifetime for VANETS. Proceedings of the 8th International Conference on ITS Telecommunications, Phuket, Thailand, 22– 24 October 2008; pp. 299–304.

50. Lee, K.; Lee, U.; Gerla, M. TO-GO: TOpology-Assist Geo-Opportunistic Routing in Urban Vehicular Grids. Proceedings of the 6th International Conference on Wireless on-Demand Network Systems and Services, Snowbird, UT, USA, 2–4 February 2009; pp. 11–18.

51. Shafiee, K.; Leung, V.C. Connectivity-aware minimum-delay geographic routing with vehicle tracking in VANETs. *Ad Hoc Netw.* **2011**, *9*, 131–141.

52. Guo, D.; Liu, Y.; Li, X.; Yang, P. False negative problem of counting bloom filter. *IEEE Trans. Knowl. Data Eng.* **2010**, *22*, 651–664.

53. Naumov, V. An Evaluation of Inter-Vehicle *Ad Hoc* Networks Based on Realistic Vehicular Traces. Proceedings of the 7th ACM International Symposium on Mobile Ad Hoc Networking and Computing (MobiHoc), Florence, Italy, 22–25 May 2006.

54. Santos, R.; Alvarez, O.; Edwards, A. Performance Evaluation of Two Location-Based Routing Protocols in Vehicular *Ad-Hoc* Networks. Proceedings of the 62nd Vehicular Technology Conference (VTC), Stockholm, Sweden, 30 May–1 June 2005.

55. Gozalvez, J.; Sepulcre, M.; Bauza, R. Impact of the radio channel modelling on the performance of VANET communication protocols. *Telecommun. Syst.* **2010**.

56. Garelli, L.; Casetti, C.; Chiasserini, C.F.; Fiore, M. MobSampling: V2V Communications for Traffic Density Estimation. Proceedings of the 73rd Vehicular Technology Conference (VTC Spring), Budapest, Hungary, 15–18 May 2011; pp. 1–5.

57. Panichpapiboon, S.; Pattara-atikom, W. Evaluation of a Neighbor-Based Vehicle Density Estimation Scheme. Proceedings of the 8th International Conference on ITS Telecommunications, Phuket, Thailand, 22–24 October 2008; pp. 294–298.

58. Molisch, A.; Tufvesson, F.; Karedal, J.; Mecklenbrauker, C. A survey on vehicle-to-vehicle propagation channels. *IEEE Wireless Commun.* **2009**, *16*, 12–22.

59. Mecklenbrauker, C.; Molisch, A.; Karedal, J.; Tufvesson, F.; Paier, A.; Bernado, L.; Zemen, T.; Klemp, O.; Czink, N. Vehicular channel characterization and its implications for wireless system design and performance. *Proc. IEEE* **2011**, *99*, 1189–1212.

60. Wang, C.; Cheng, X.; Laurenson, D. Vehicle-to-vehicle channel modeling and measurements: Recent advances and future challenges. *IEEE Commun. Mag.* **2009**, *47*, 96–103.

61. Acosta-Marum, G.; Ingram, M. Six Time- and Frequency-Selective Empirical Channel Models for Vehicular Wireless LANs. Proceedings of the 66th Vehicular Technology Conference, Baltimore, MD, USA, 30 September–3 October 2007; pp. 2134–2138.

62. Cheng, L.; Henty, B.; Bai, F.; Stancil, D. Highway and Rural Propagation Channel Modeling for Vehicle-to-Vehicle Communications at 5.9 GHz. Proceedings of the 2008 IEEE Antennas and Propagation Society International Symposium, San Diego, CA, USA, 5– 11 July 2008; pp. 1–4.

63. Kunisch, J.; Pamp, J. Wideband Car-to-Car Radio Channel Measurements and Model at 5.9 GHz. Proceedings of the IEEE 68th Vehicular Technology Conference, Calgary, Alberta, 21– 24 September 2008; pp. 1–5.

64. Paier, A.; Karedal, J.; Czink, N.; Dumard, C.; Zemen, T.; Tufvesson, F.; Molisch, A.; Mecklenbruker, C. Characterization of vehicle-to-vehicle radio channels from measurements at 5.2GHz. *Wireless Pers. Commun.* **2009**, *50*, 19–32.

65. Zrar Ghafoor, K.; AbuBakar, K.; van Eenennaam, M.; Khokhar, R.; Gonzalez, A. A fuzzy logic approach to beaconing for vehicular *ad hoc* networks. *Telecommun. Syst.* **2011**.

66. Treiber, M.; Hennecke, A.; Helbing, D. Congested traffic states in empirical observations and microscopic simulations. *Phys. Rev. E* **2000**, *62*, 1805–1824.

67. Wang, H.; Ni, D.; Chen, Q.Y.; Li, J. Stochastic modeling of the equilibrium speeddensity relationship. *J. Adv. Transp.* **2011**.

68. Behnad, A.; Nader-Esfahani, S. On the statistics of MFR routing in one-dimensional *ad hoc* networks. *IEEE Trans. Veh. Tech.* **2011**, *60*, 3276–3289.

69. Opnet Modeler 16.0. Available online: http://www.opnet.com/solutions/network_rd/modeler.html(accessed on 15 September 2012).

70. Okamura, T.; Ideguchi, T.; Tian, X.; Okuda, T. Traffic Evaluation of Group Communication Mechanism among Vehicles. Proceedings of the 4th International Conference on Computer Sciences and Convergence Information Technology, Seoul, Korea, 24–26 November 2009.

71. Opnet Contributed Papers. Avaliable online: https://enterprise1.opnet.com/tsts/4dcgi/BiblioSearchSubmit?QueryBiblio_whatFindAll&QueryRecordsPerPage1500 (accessed on 8 October 2012).

72. Dressler, F.; Sommer, C.; Eckhoff, D.; Tonguz, O. Toward Realistic Simulation of Intervehicle Communication. *IEEE Veh. Tech. Mag.* **2011**, *6*, 43–51.

CITATION

Gabriel A. Galaviz-Mosqueda, Raúl Aquino-Santos, Salvador Villarreal-Reyes, Raúl Rivera-Rodríguez, Luis Villaseñor-González and Arthur Edwards , Reliable Freestanding Position-Based Routing in Highway Scenarios, doi:10.3390/s121114262.

Index

A

a Geographic Information
System (GIS), 183
Accident, 43, 46, 47, 49, 50, 58
accident forecast model, 43
Analytic minimum impedance
surface (AMIS), 159
automated traffic measuring
system, 95
autonomous vehicle, 61, 64, 68,
91, 92
Average delay, 147, 148

C

Cellular automaton (CA), 2
Center for Transport and
Logistics (CLT), 189
Communication Access for Land
Mobile (CALM), 228
Competitive effect, 149
comprehensive model, 3
Congestion impedance, 172
control strategy, 117, 131, 132

D

decision making, 61, 66, 92
Dedicated Short-Range
Communications (DSRC), 228
Demographic density, 188, 191,
199, 202
density, 1, 2, 3, 10, 11, 12, 14, 15,
16, 17, 18
detect incident, 43
detection algorithm, 43, 44
digital age, 43
Distribution function, 145, 146
Driving hazard condition, 159

E

Ecological study, 191
Economical transport, 179
Endogeneity, 187
Exponential distribution, 142,
144

F

Fargo-Moorhead metropolitan,
164
First-in first-out (FIFO), 158

Fixed Time Ramp Metering
(FTRM), 117
Freestanding Position Based
Routing (FPBR), 230, 235
Freestanding position-based
routing algorithm (FPBR), 227

G

Geographic information systems
(GIS), 159
Geographic Information Systems
(GIS), 155
GIS network, 157
Graphical User Interface, 155

H

high maintenance cost, 96, 97,
113
high speed limit, 44
High vehicle mobility, 230, 231
Highway scenario, 227, 228, 229,
230, 231, 232, 233, 234, 235,
236, 237, 247, 249, 251, 253, 261
human-driven vehicle, 61

I

indicator, 117, 131, 132
Intelligent Transport Systems
(ITS), 228
intelligent transportation
system, 95
Intersection delay, 138, 139, 140,
148, 149, 152
Intersection traffic delay, 137

K

Kernel density estimation (KDE),
180, 182

L

Left-turning vehicles, 137, 152
Level of service (LOS), 138
low error rate, 95, 111

M

Medium Access Control (MAC),
229, 263
microscopic model, 2, 3
Mimics stochastic, 172
Mixed traffic flow, 1, 2
motion planning, 61, 63, 65, 66,
67, 93
Motorized transportation, 180,
181, 183, 203
multiple drive lane, 95

N

Neighborhood matrix, 187, 192
Non-motorized transportation,
179

O

Ordinary least squares (OLS),
180, 187
overtaking, 61, 62, 63, 64, 65, 66,
67, 69, 70, 71, 72, 73, 74, 75, 76,
77, 79, 80, 81, 82, 83, 85, 86, 87,
88, 90, 93

P

Pedestrian crossing, 142, 143
Pertinent parameter, 155

planning algorithm, 62
planning framework, 62, 90
Poisson distribution, 142, 143,
    144
proportion of bicycle, 1, 14, 16

R

ramp metering, 117, 118, 119,
    122, 123, 124, 127, 129, 131, 132
real traffic road, 95
Realistic traffic assignment, 155
Retransmission mechanism, 230,
    232
Right-turn traffic delay, 137, 147
Right-turning traffic, 138, 139,
    149
roadway environment, 95, 96
Routing mechanism, 230
Routing strategy, 229

S

scanning method, 95, 96
Shift-share analysis, 138, 140,
    148, 152, 154
Shift-share analysis (SSA), 138
single equity measure, 117
Socio-economic conditions, 185
speed lane, 61, 62, 63, 91, 92
Statistical analysis, 191, 201, 202,
    204

T

Traffic accident risk, 183
Traffic assignment model, 156,
    157
Traffic assignment model
    (TAM), 156

traffic condition, 44
Traffic delay, 139
traffic flow simulation software,
    117, 127
traffic management strategy, 117
Traffic measurement
    technology, 96
traffic scenario, 61, 93
traffic volume, 95, 96, 110
transportation infrastructure, 44
Transportation research, 182
Transportation science, 181
Transportation system, 180, 181,
    182, 183, 185
Travers ability, 161, 162, 163, 164,
    166, 173, 175
Trip generator hubs (TGH), 180,
    191

U

Urban transportation system, 179

V

variation, 1, 14, 16
vehicle, 1, 2, 3, 4, 5, 6, 7, 8, 9, 11,
    14, 15, 16, 17, 18, 20
Vehicle communication, 228, 230,
    265
vehicles and drivers, 44
velocity, 1, 3, 4, 5, 6, 7, 8, 10, 11,
    12, 14, 16, 18
Vulnerable road users (VRU),
    180, 181, 204

W

World Health Organization
    (WHO), 183, 205, 206